Modified Method for Estimating Petroleum Source-Rock Potential Using Wireline Logs, With Application to the Kingak Shale, Alaska North Slope

Scientific Investigations Report 2016–5001

U.S. Department of the Interior
U.S. Geological Survey

Cover. Outcrop along Hue Creek at the northern front of the Shublik Mountains, northeast Brooks Range. A geologist examines the steeply dipping, overturned contact between the Jurassic Kingak Shale (left-center) and Lower Cretaceous Kemik Sandstone (right). The light-colored rocks (upper left) are Proterozoic Katakturuk Dolomite, thrust northward over the Kingak Shale. The Kingak-Kemik contact is the Lower Cretaceous unconformity. Photograph by David W. Houseknecht, U.S. Geological Survey.

Modified Method for Estimating Petroleum Source-Rock Potential Using Wireline Logs, With Application to the Kingak Shale, Alaska North Slope

By William A. Rouse and David W. Houseknecht

Scientific Investigations Report 2016–5001

U.S. Department of the Interior
U.S. Geological Survey

U.S. Department of the Interior
SALLY JEWELL, Secretary

U.S. Geological Survey
Suzette M. Kimball, Director

U.S. Geological Survey, Reston, Virginia: 2016

For more information on the USGS—the Federal source for science about the Earth, its natural and living resources, natural hazards, and the environment—visit http://www.usgs.gov or call 1–888–ASK–USGS.

For an overview of USGS information products, including maps, imagery, and publications, visit http://www.usgs.gov/pubprod/.

Suggested citation:
Rouse, W.A., and Houseknecht, D.W., 2016, Modified method for estimating petroleum source-rock potential using wireline logs, with application to the Kingak Shale, Alaska North Slope: U.S. Geological Survey Scientific Investigations Report 2016–5001, 40 p., http://dx.doi.org/10.3133/sir20165001.

ISSN 2328-0328 (online)

Contents

Figures

Conversion Factors

Multiply	By	To obtain
	Length	
foot (ft)	0.3048	meter (m)
mile (mi)	1.609	kilometer (km)

Datum

Horizontal coordinate information is referenced to the North American Datum of 1927 (NAD 27).

Abbreviations

API	American Petroleum Institute
$\Delta log\ R$	delta-log resistivity
DT	sonic travel time
GR	gamma-ray
HGR	high-gamma-ray
LAS	Log ASCII Standard
LCU	Lower Cretaceous unconformity
μsec/ft	microseconds per foot
NPRA	National Petroleum Reserve in Alaska
ohm-m	ohm-meters
RILD	resistivity
RMA	reduced major axis
TOC	total organic carbon
USGS	U.S. Geological Survey

Modified Method for Estimating Petroleum Source-Rock Potential Using Wireline Logs, With Application to the Kingak Shale, Alaska North Slope

By William A. Rouse and David W. Houseknecht

Abstract

In 2012, the U.S. Geological Survey completed an assessment of undiscovered, technically recoverable oil and gas resources in three source rocks of the Alaska North Slope, including the lower part of the Jurassic to Lower Cretaceous Kingak Shale. In order to identify organic shale potential in the absence of a robust geochemical dataset from the lower Kingak Shale, we introduce two quantitative parameters, $\Delta DT_{\bar{x}}$ and ΔDT_z, estimated from wireline logs from exploration wells and based in part on the commonly used delta-log resistivity ($\Delta log\ R$) technique. Calculation of $\Delta DT_{\bar{x}}$ and ΔDT_z is intended to produce objective parameters that may be proportional to the quality and volume, respectively, of potential source rocks penetrated by a well and that can be used as mapping parameters to convey the spatial distribution of source-rock potential. Both the $\Delta DT_{\bar{x}}$ and ΔDT_z mapping parameters show increased source-rock potential from north to south across the North Slope, with the largest values at the toe of clinoforms in the lower Kingak Shale. Because thermal maturity is not considered in the calculation of $\Delta DT_{\bar{x}}$ or ΔDT_z, total organic carbon values for individual wells cannot be calculated on the basis of $\Delta DT_{\bar{x}}$ or ΔDT_z alone. Therefore, the $\Delta DT_{\bar{x}}$ and ΔDT_z mapping parameters should be viewed as first-step reconnaissance tools for identifying source-rock potential.

Introduction

In 2012, the U.S. Geological Survey (USGS) completed an assessment of undiscovered, technically recoverable oil and gas resources in three source-rock systems (fig. 1) of the Alaska North Slope: (1) the Triassic Shublik Formation; (2) the lower part of the Jurassic to Lower Cretaceous Kingak Shale; and (3) the Cretaceous pebble shale unit, Hue Shale, and parts of the Paleogene Canning Formation, collectively called the Brookian shale (Houseknecht, Rouse, Garrity, and others, 2012). Maps of inferred source-rock richness were constructed using three parameters because of differences in lithology and wireline-log response among the source rocks. The map used for the Kingak Shale is highly generalized (fig. 2) because no quantitative mapping parameter had been defined. The study summarized in this report was initiated to evaluate the efficacy of the delta-log resistivity ($\Delta log\ R$) technique (Passey and others, 1990) for estimating an objective and quantitative parameter for evaluating source-rock potential from wireline-log data. This parameter may be useful as an evaluation tool for individual wells and, when calculated for multiple wells, as a mapping parameter. However, due to software limitations for digital calculation of $\Delta log\ R$ and the absence of a robust geochemical dataset for calibration of $\Delta log\ R$, we sought to develop a modified version of $\Delta log\ R$ that yields two parameters that may serve as proxies of source-rock quality and volume. This report documents the digital workflow developed for calculating a modified version of $\Delta log\ R$ and presents the results of applying the technique to evaluate source-rock potential of the lower Kingak Shale.

Geologic Background

The Jurassic to Lower Cretaceous Kingak Shale contains both marine and terrigenous organic matter deposited in a marine siliciclastic setting influenced by pulses of syndepositional uplift of the Beaufort rift shoulder (also known as the Barrow arch; see figs. 3 and 4) during opening of the Canada Basin (Magoon and Claypool, 1984; Hubbard and others, 1987; Bird and Houseknecht, 2011). Houseknecht and Bird (2004) identified four depositional sequence sets in the Kingak Shale (fig. 3, K1–K4) that define a northern-sourced southward-offlapping succession of Beaufortian strata in the National Petroleum Reserve in Alaska (NPRA), and these sequence sets subsequently have been mapped eastward beyond the NPRA on the basis of seismic and well data.

The basal K1 sequence set is 1,000 to more than 1,250 feet (ft) (300 to 380 meters [m]) thick across a broad area in the north-central NPRA that extends to the south as a lobe in the central NPRA (fig. 4). North of the zone of

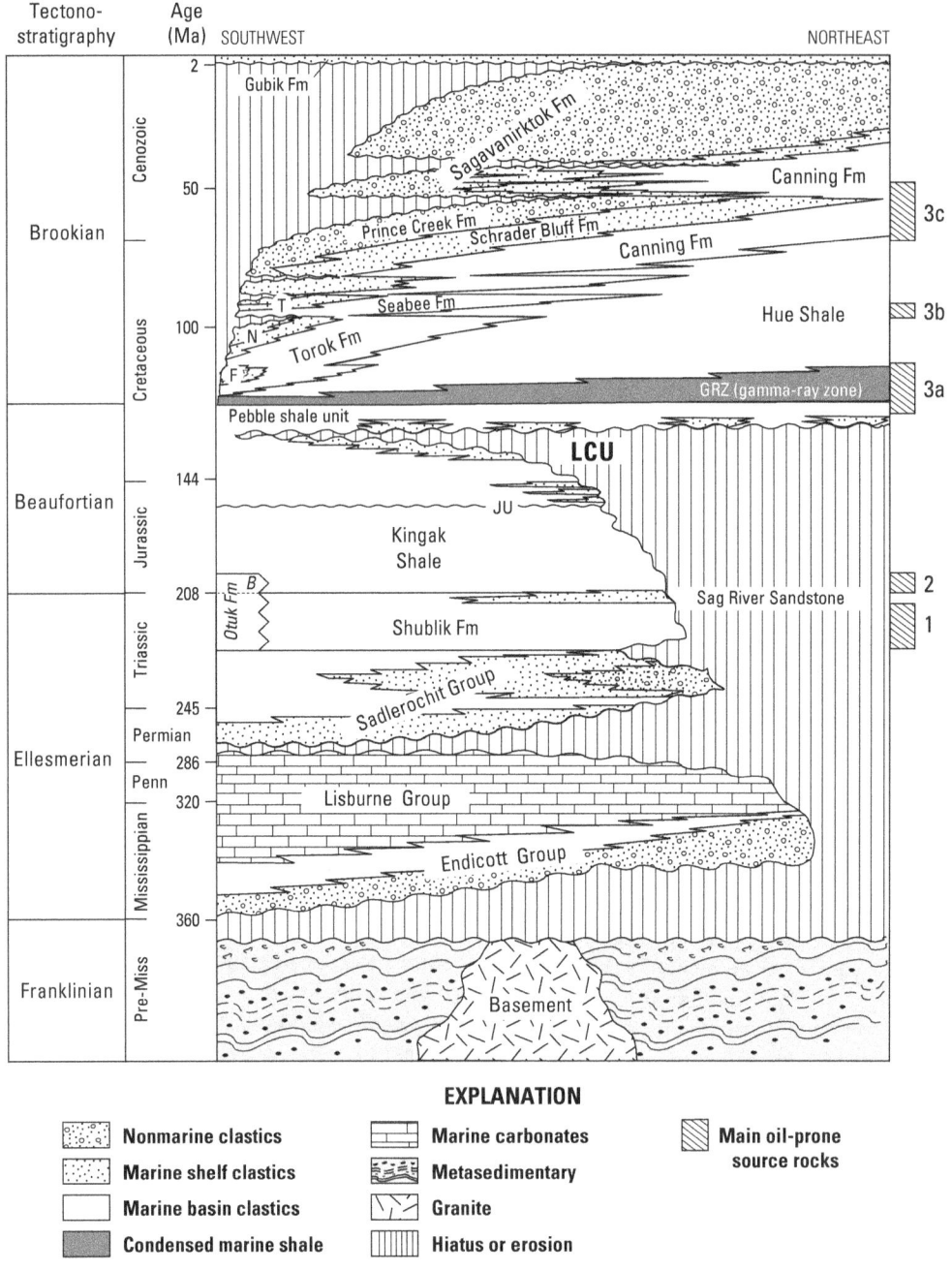

Figure 1.   Diagram showing generalized chronostratigraphy for the Alaska North Slope (from Houseknecht, Bird, and Garrity, 2012). Oil-prone source-rock systems discussed in text are indicated at right: 1, Triassic source-rock system, composed of the Shublik Formation and the Triassic part of the Otuk Formation; 2, Jurassic source-rock system, composed of the lower part of the Kingak Shale and the Blankenship Member (B) of the Otuk Formation; and 3, Cretaceous to Cenozoic source-rock system, composed of (a) the Lower Cretaceous pebble shale unit and gamma-ray zone (GRZ), (b) Upper Cretaceous organic-matter-rich tongues of the Hue Shale, and (c) lower Paleogene organic-matter-rich tongues of the Canning Formation. Italicized labels (Otuk Formation and B) indicate units that crop out in the Brooks Range frontal thrust belt and that represent southern distal facies equivalents of formations present beneath the Alaska North Slope. Arctic Alaska stratigraphy modified from Lerand (1973), Bird (1985, 2001), Hubbard and others (1987), and Mull and others (2003); ages from Gradstein and others (2004) Abbreviations used: B, Blankenship Member of the Otuk Formation; F, Fortress Mountain Formation; Fm., Formation; GRZ, gamma-ray zone; JU, Jurassic unconformity; LCU, Lower Cretaceous unconformity; Ma, mega-annum, or million years ago; N, Nanushuk Formation; PENN, Pennsylvanian; Perm, Permian; Pre-Miss, pre-Mississippian; T, Tuluvak Formation.

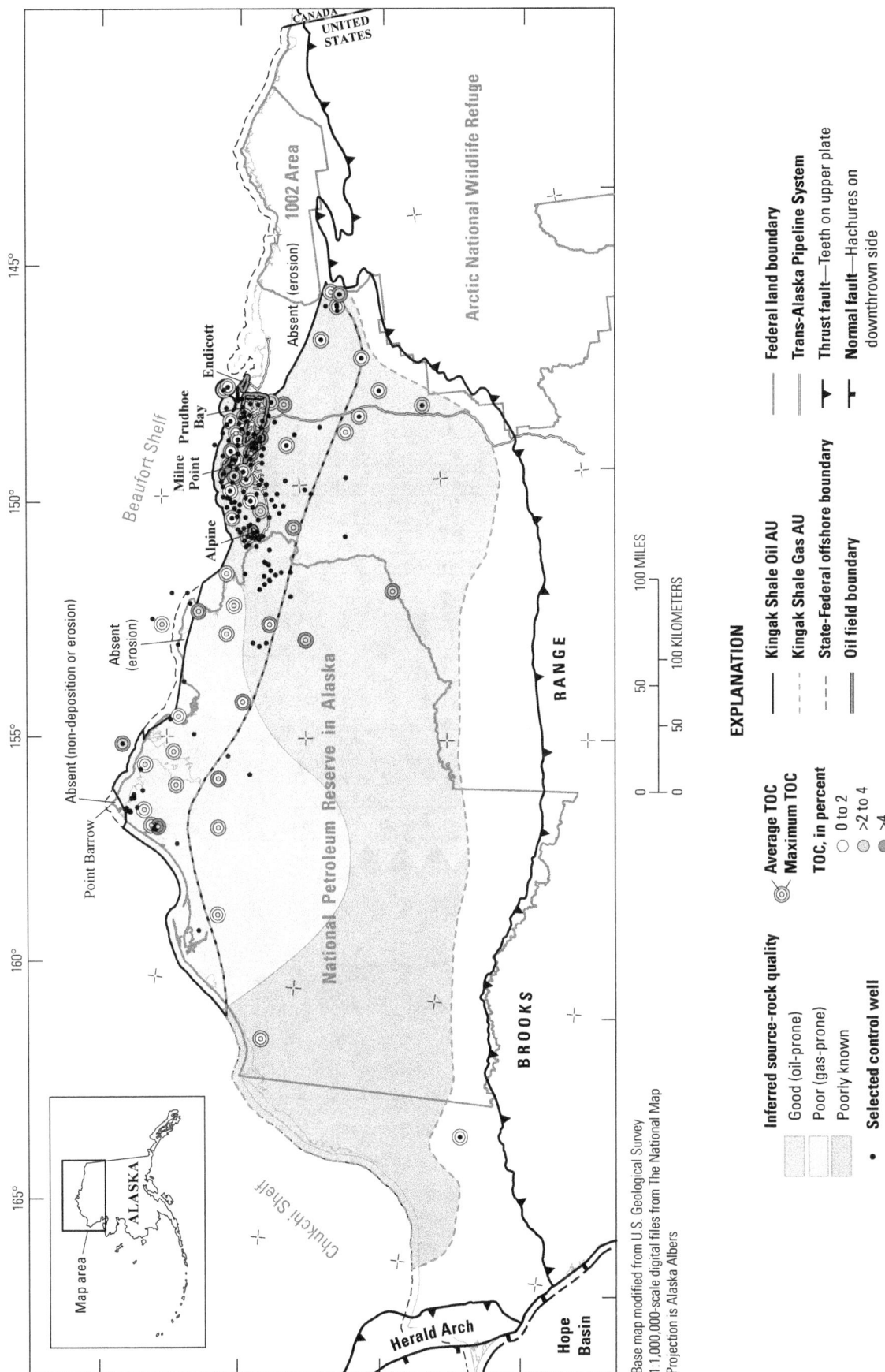

Figure 2. Map of northern Alaska showing areas of inferred source-rock quality in the Kingak Shale and outlines of shale-oil and shale-gas assessment units (AUs). Average and maximum total organic carbon (TOC) content of the Kingak Shale from public-domain data (Threlkeld and others, 2000; Peters and others, 2006). Areas of inferred source-rock quality south of well control are based on sequence stratigraphic interpretations (Houseknecht and Bird, 2004), seismic observations, and outcrops of the Kingak Shale and coeval rocks in the Brooks Range foothills. Note that the source-rock-quality map does not extend beyond the State-Federal offshore boundary and that only selected other information is shown beyond that boundary. Modified from Houseknecht, Rouse, and Garrity (2012).

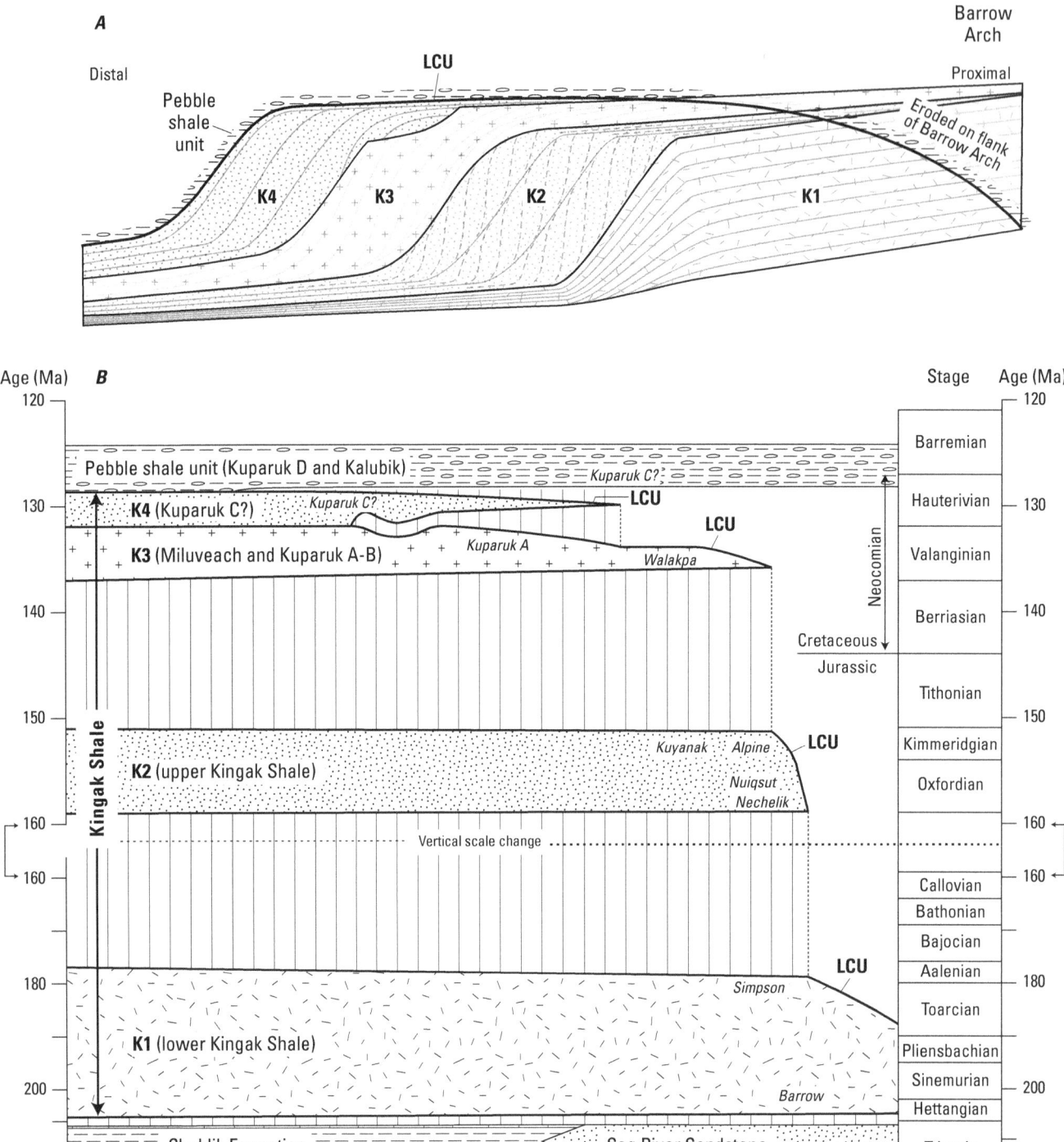

Figure 3. Schematic summary of inferred depositional sequence sets in the Kingak Shale in the National Petroleum Reserve in Alaska (modified from Houseknecht and Bird, 2004). *A*, Cross section from south (distal) to north (proximal) showing lithostratigraphic relations among four depositional sequence sets (K1–K4) and general aspects of internal stratal geometry. Note that the Lower Cretaceous unconformity (LCU) bevels Kingak Shale strata northward to extinction on the Barrow arch. *B*, Diagram showing chronostratigraphy of sequence sets in the Kingak Shale. Shown in parentheses are stratigraphic names widely applied to approximate age-equivalent strata of the four sequence sets. Names in italics show approximate positions of sandstone reservoirs and potential reservoirs. Vertical lines denote time gaps between sets. Note scale change near center of diagram. Geologic timescale from Gradstein and Ogg (1996). Abbreviation used: Ma, mega-annum, or million years ago.

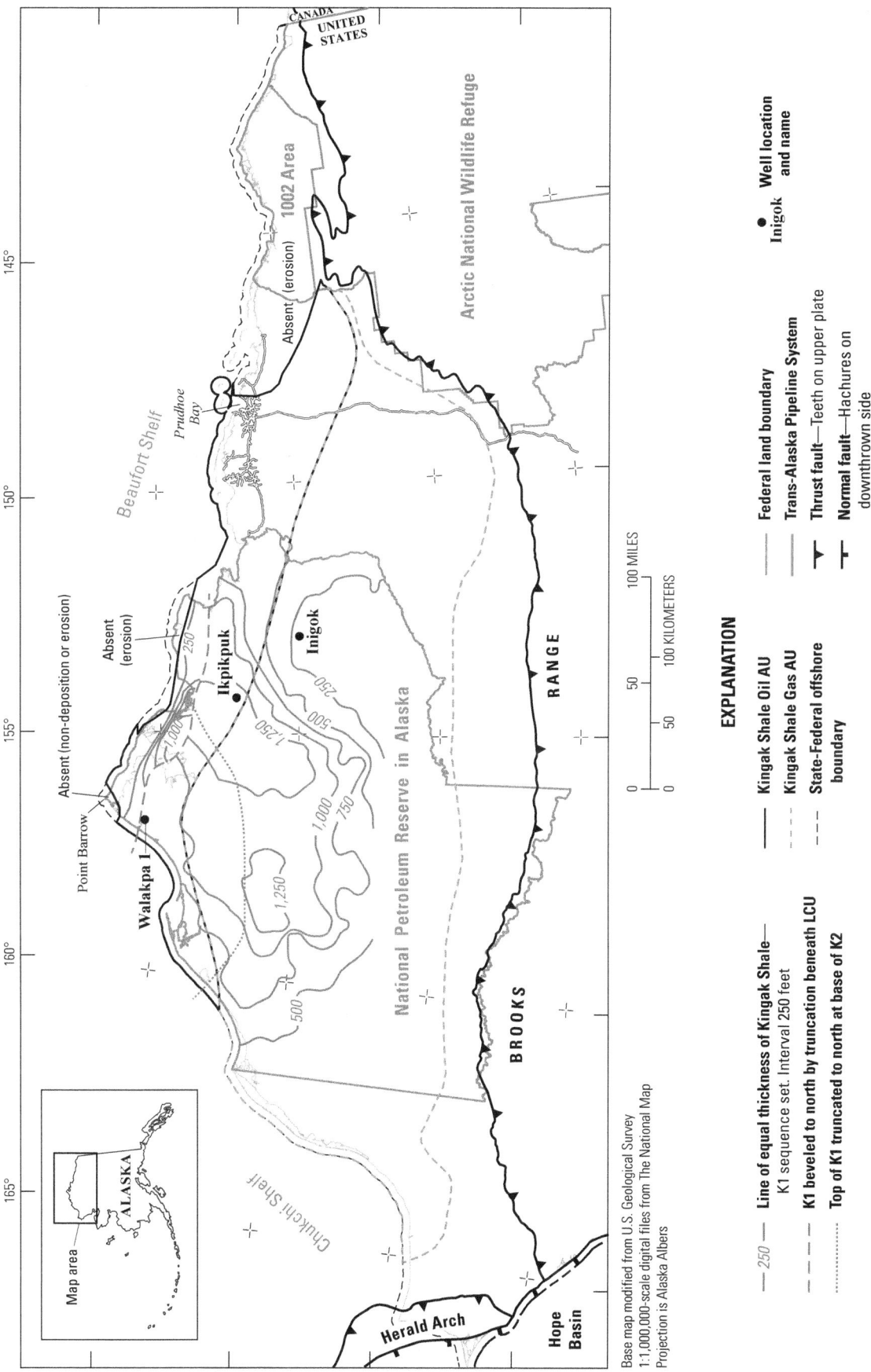

Base map modified from U.S. Geological Survey
1:1,000,000-scale digital files from The National Map
Projection is Alaska Albers

EXPLANATION

— *250* —   **Line of equal thickness of Kingak Shale—**
        K1 sequence set. Interval 250 feet

— — —   **K1 beveled to north by truncation beneath LCU**

·········   **Top of K1 truncated to north at base of K2**

———   **Federal land boundary**

— — —   **Trans-Alaska Pipeline System**

— — —   **State-Federal offshore
        boundary**

        Kingak Shale Oil AU

        Kingak Shale Gas AU

▼   **Thrust fault**—Teeth on upper plate

⊥   **Normal fault**—Hachures on
        downthrown side

●   **Well location
Inigok   and name**

Figure 4. Map of the National Petroleum Reserve in Alaska (NPRA) showing thickness of Kingak Shale sequence set K1. Modified from Houseknecht and Bird (2004). Abbreviations used: AU, assessment unit; LCU, Lower Cretaceous unconformity.

maximum thickness in the central NPRA, the K1 sequence set thins gradually where seismic data show an erosional contact between the K1 and K2 sequence sets. Farther north, the K1 sequence set abruptly thins to a zero edge along the trend of the Barrow arch, where it is beveled beneath the Lower Cretaceous unconformity. South of the zone of maximum thickness, the K1 sequence set abruptly thins in a radial pattern to less than 500 ft (150 m) in the western NPRA and to less than 250 ft (76 m) in the southeastern NPRA.

Wireline-log responses within the K1 sequence set in the northern NPRA commonly exhibit thin (<200 ft [61 m]) coarsening-upward trends consisting of mudstone grading upward to siltstone (for example, fig. 5, Walakpa 1 well, 2,980- to 2,800-ft [908- to 853-m] depth) or mudstone grading upward to sandstone (for example, fig. 5, Walakpa 1 well, 3,200- to 3,050-ft [975- to 930-m] depth). In contrast, fining-upward transitions commonly are abrupt between sandstone and siltstone (for example, fig. 5, Walakpa 1 well, ~3,050-ft [930-m] depth) or between siltstone and mudstone (for example, fig. 5, Walakpa 1 well, ~2,990-ft [911-m] depth), with an additional few thin fining-upward successions also present (for example, fig. 5, Walakpa 1 well, 2,800- to 2,770-ft [853- to 844-m] depth; Houseknecht and Bird, 2004).

Within the zone of maximum thickness in the central NPRA, wireline-log responses within the K1 sequence set display subtle coarsening-upward successions capped by siltstone abruptly overlain by shale, as well as repetitive intervals of silty mudstone locally punctuated by shale (fig. 5, Ikpikpuk well). The wireline-log response in the most distal well penetrations of the K1 sequence set in the eastern NPRA displays an off-the-scale gamma-ray response in a thin interval of silty mudstone near or beyond the toe of K1 clino-forms (fig. 5, Inigok well).

The distal increase in gamma-ray response within the K1 sequence set was interpreted as increased organic matter content and coalescence of time lines (that is, it is a condensed section) in a distal direction by Houseknecht and Bird (2004). Biostratigraphic data from NPRA wells indicate that the K1 sequence set is Early to Middle Jurassic, equivalent to the lower Kingak Shale in the central North Slope (Carman and Hardwick, 1983; Masterson and Paris, 1987). Conventional oil accumulations at the Alpine, Endicott, Milne Point, and Prudhoe Bay fields were sourced entirely or partly from the lower Kingak Shale (fig. 2; Seifert and others, 1980; Claypool and Magoon, 1985; Premuzic and others, 1986; Sedivy and others, 1987; Masterson, 2001; Magoon and others, 2003; Peters and others, 2006). The potential for self-sourced, continuous accumulations of recoverable oil and gas in lower Kingak Shale source rocks is assumed to exist (Houseknecht, Rouse, and Garrity, 2012) but has not been confirmed by well completions.

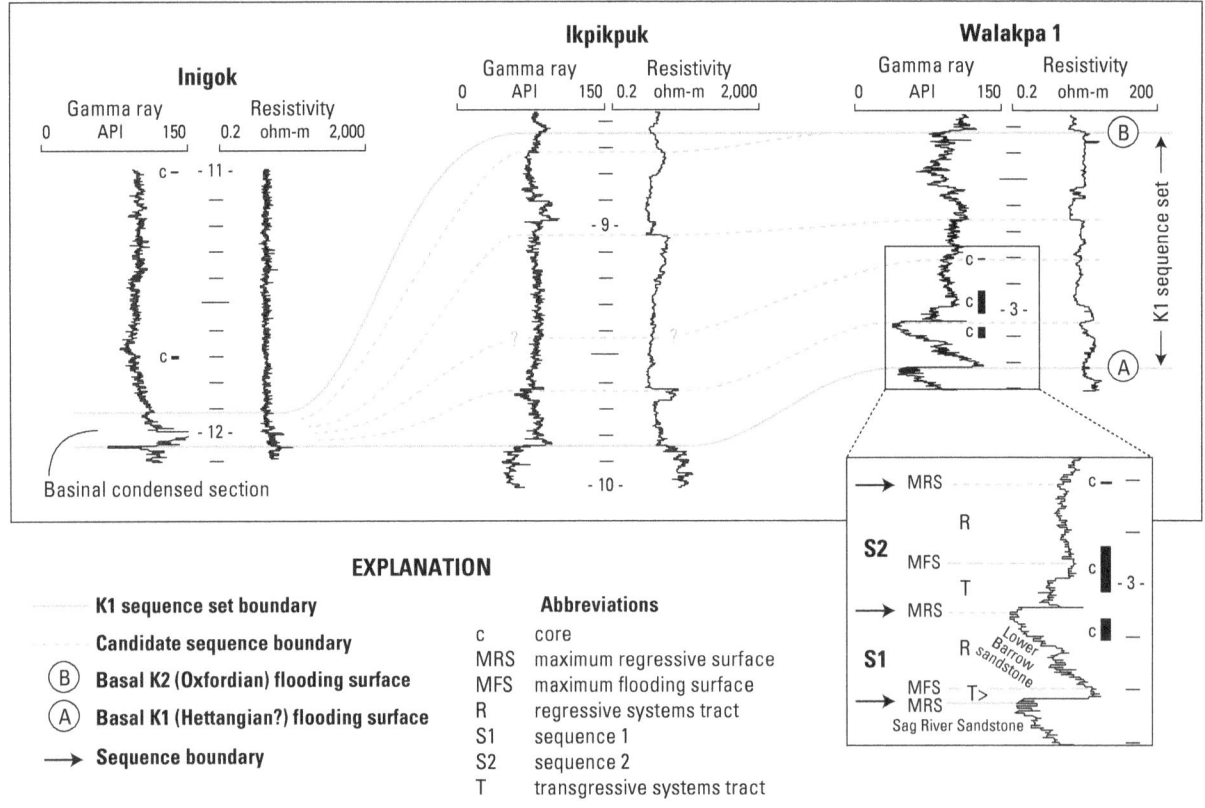

Figure 5. Wireline-log response in the K1 sequence set. Well locations are shown in figure 4. Wireline-log measured depth ticks below kelly bushing are at 100-foot (ft) (30-meter) intervals, and numbers are ×1,000 ft. Modified from Houseknecht and Bird (2004). Abbreviations used: ohm-m, ohm-meter; API, American Petroleum Institute.

Source-Rock Characterization With Wireline Logs

Petroleum source rocks typically are shale or limestone containing more than 1 or 2 weight percent of organic matter (Tissot and Welte, 1984). Direct geochemical measurements on source rocks are generally sparse, resulting in the increased use of common wireline logs from exploration and development wells for identifying source-rock intervals and estimating organic matter content. Recognition of organic-matter-rich strata from wireline logs is based on the unique physical properties of organic matter as compared to minerals in the host rock. These properties include higher radioactivity (Beers, 1945; Schmoker, 1981), lower density (Schmoker, 1979), higher resistivity (Nixon, 1973; Meissner, 1978; Schmoker and Hester, 1989), and slower sonic velocity or higher sonic travel time (Dellenbach and others, 1983).

Previous assessments of technically recoverable shale-gas resources by the USGS have used a high-gamma-ray (HGR; gamma-ray greater than 150 American Petroleum Institute [API] units) mapping parameter as a possible indication of source-rock richness (Houseknecht and others, 2014). Whereas gamma-ray response increases distally within the lower Kingak Shale, gamma-ray values rarely exceed 150 API except for a thin interval near the base of the formation in distal parts of the depositional system, precluding the use of the HGR mapping parameter in identifying source-rock potential and necessitating an alternative methodology.

Meyer and Nederlof (1984) developed a method involving a combination of density, resistivity, and sonic logs that discriminates between source rocks and non-source rocks without attempting to quantify the organic-matter richness from the combination of logs. Their technique uses cross plots of density versus resistivity and of sonic travel time versus resistivity; strata with relatively high resistivity and either relatively high sonic travel time or low bulk density represent a potential source rock. A regression line is fit through the cross-plot data, the equation of which becomes the discriminant function for separating potential source rock from non-source rock.

Passey and others (1990), using a principle similar to that of Meyer and Nederlof (1984), developed a method called delta-log resistivity ($\Delta log\ R$) that identifies potential source rocks by overlaying the sonic curve and the resistivity curve in a baseline interval consisting of clay-rich rocks (mudstone or shale) that are not of source-rock quality (fig. 6). Potential source rocks in other depth intervals of the well are identified by a separation of the two curves through the parameter quantified in the following equation:

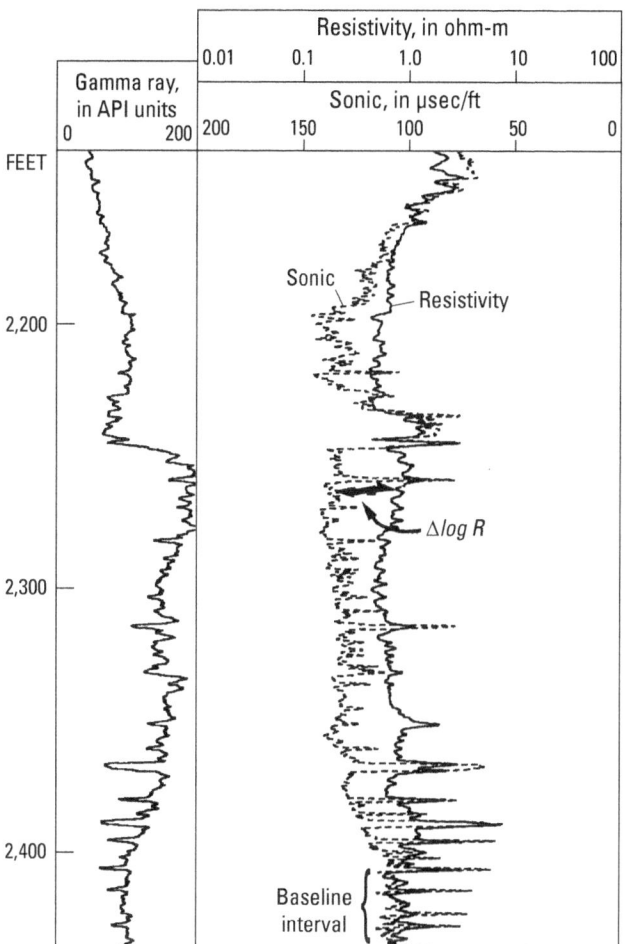

Figure 6. Part of a wireline log illustrating overlay of sonic and resistivity logs to define $\Delta log\ R$ separation in an unidentified organic-matter-rich interval. Scaling of the sonic and resistivity curves is adjusted so that 50 μsec/ft on the sonic log corresponds to one decade of resistivity. The values in the center of the sonic and resistivity log track correspond to the $R_{baseline}$ and $\Delta t_{baseline}$ values (for this example, $R_{baseline}$ = 1 ohm-m, and $\Delta t_{baseline}$ = 100 μsec/ft). Wireline-log measured depth ticks below kelly bushing are at 100-foot (ft) (30-meter) intervals. Figure from Passey and others (1990). Abbreviations used: ohm-m, ohm-meter; μsec/ft, microsecond per foot; m, meter; API, American Petroleum Institute.

$$\Delta log\,R = log_{10}\left(\frac{R}{R_{baseline}}\right) + 0.02 \times (\Delta t - \Delta t_{baseline}) \quad (1)$$

where

$\Delta log\,R$	is the curve separation measured in logarithmic resistivity cycles;
R	is the resistivity measured in ohm-meters (ohm-m);
Δt	is the measured sonic travel time in microseconds per foot (µsec/ft);
$R_{baseline}$	is the resistivity corresponding to the $\Delta t_{baseline}$ value when the curves are overlain in non-source, clay-rich rocks; and
0.02	is based on the ratio of -50 µsec/ft per resistivity cycle.

Passey and others (1990) found a linear correlation between $\Delta log\,R$ separation and total organic carbon (TOC) content in multiple source rocks as a function of thermal maturity (fig. 7). The original calibration of the $\Delta log\,R$ technique (Passey and others, 1990) was for source rocks in the oil window, as there was no calibration at that time to include rocks of higher thermal maturity (Passey and others, 2010). Sondergeld and others (2010) proposed using a correction multiplier to obtain log-derived TOC using the $\Delta log\,R$ technique for overmature shale-gas formations:

$$TOC = \Delta log\,R \times 10^{(2.297 - 0.1688 \times LOM)} \times C \quad (2)$$

where

TOC	is the total organic carbon measured in weight percent,
LOM	is the level of organic metamorphism (Hood and others, 1975), and
C	is a correction factor.

Methodology

The methodology used in this study for identifying organic shale potential is based on a combination of the cross-plot and $\Delta log\,R$ methods (Meyer and Nederlof, 1984; Passey and others, 1990) and follows an example presented by Bowman (2010). Our digital workflow was developed and tested using IHS Kingdom® version 8.8 software. The following procedures were performed on a well-by-well basis.

To assure adequate non-source-rock strata with which to determine a baseline for sonic and resistivity data, we defined a target stratigraphic interval that includes both the K1 and the overlying K2 sequence sets (fig. 3) in the Kingak Shale, where the K2 sequence set is assumed to consist of predominately non-source-rock strata. The target stratigraphic interval was identified by examining wireline-log and seismic data across the North Slope. To constrain the analysis to clay-rich intervals within the K2 and K1 sequence sets, shale volume was calculated for each well using the following equation:

$$V_{sh} = \left(\frac{GR_{log} - GR_{clean}}{GR_{shale} - GR_{clean}}\right) \quad (3)$$

where

V_{sh}	is the shale volume (decimal percent),
GR_{log}	is the API value from the gamma-ray log curve,
GR_{clean}	is the API value of a "clean" sand, and
GR_{shale}	is the API value of a shale.

GR_{clean} and GR_{shale} values were computed for each well using the IHS Kingdom® version 8.8 software Petrophysics module. Only strata consisting of at least 60 percent shale

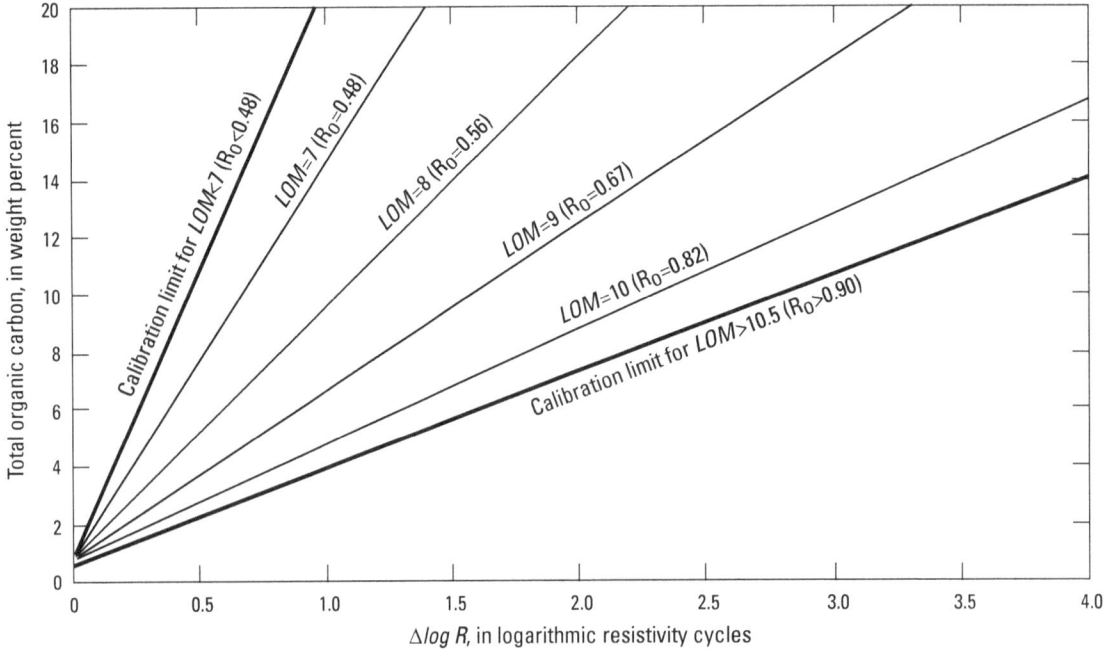

Figure 7.   Chart relating $\Delta log\,R$, total organic carbon, and thermal maturity expressed as level of organic metamorphism (*LOM*) (Hood and others, 1975). Abbreviations used: R$_o$, vitrinite reflectance.

by volume ($V_{sh} \geq 0.6$) were considered suitable as a baseline (Tom Wild, President and Owner, Tom Wild Petrophysical Services LLC, written commun., February 27, 2013). Fifty-four wells were identified with suitable shale intervals and complete digital gamma-ray (GR), resistivity (RILD), and sonic travel time (*DT*) wireline-log data that extend through the K2 and K1 sequence sets.

The inability to display linear and logarithmic data on the same log track in IHS Kingdom® version 8.8 software precluded digital curve manipulation of the $\Delta log\ R$ methodology (Passey and others, 1990) to determine baseline values for non-source-rock shale intervals. As an alternative, cross plots of sonic travel time versus resistivity were constructed for each well (Bowman, 2010). Cross-plot data were constrained to the assumed non-source-rock K2 sequence set and a reduced major axis (RMA) regression line was fit through the data, thus automating the determination of Bowman's (2010) low-resistivity shale line. The resultant correlation equation then was used to calculate a pseudosonic log DT_{logR} which transformed resistivity data into sonic travel time units (μsec/ft), thereby enabling the direct comparison of sonic and resistivity log data within the same unit space and scaling the resistivity data to overlie the sonic data in the assumed non-source-rock interval (equation 4, fig. 8).

$$DT_{logR} = b - m \times log\ R \qquad (4)$$

where

    m   is the slope and
    b   is the y-intercept of a line.

In two cases where individual wells exhibited an inverse regression trend (positive *m* value), values for *b* and *m* were substituted from the nearest well. The DT_{logR} curve calculation (equation 4) then was applied to both K2 and K1 sequence sets. Curve separation within the lower Kingak Shale K1 sequence set was calculated using the following equation:

$$\Delta DT = DT - DT_{logR} \qquad (5)$$

where ΔDT is the separation between *DT* and DT_{logR} curves (in units of μsec/ft), a transform functionally similar to $\Delta log\ R$ of Passey and others (1990).

It should be noted that whereas ΔDT is functionally similar to $\Delta log\ R$ of Passey and others (1990), the application of the RMA transform in the calculation of DT_{logR} ultimately distorts the resistivity data. Thus, ΔDT values may not be equivalent to $\Delta log\ R$ values in all cases.

We assumed the existence of a positive relationship between $\Delta log\ R$ and TOC, as documented by Passey and others (1990, 2010) (fig. 7), to infer qualitatively the presence of potential source-rock intervals in the lower Kingak Shale K1 sequence set. Although this revised methodology addressed the presence of potential source rocks by the calculation of ΔDT, our objectives also included the development of a parameter that may be proportional to the volume of potential source rocks in a well. We therefore introduced an additional

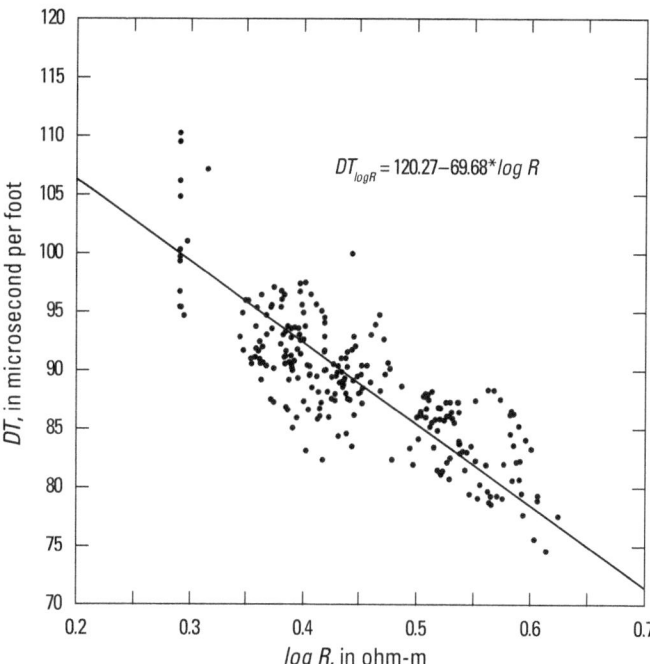

Figure 8. Cross plot of log resistivity (*log R*) versus sonic travel time (*DT*) (both derived from wireline logs) within the K2 sequence set of the Kingak Shale in the Ikpikpuk No. 1 well (location shown in figures 9 and 10). The equation of the reduced major axis (RMA) regression line is used to transform resistivity-log data into sonic travel time units, enabling the direct comparison of sonic- and resistivity-log data within the same unit space. Abbreviations used: ohm-m, ohm-meter; μsec/ft, microsecond per foot.

parameter, ΔDT_z, which incorporated both the magnitude and thickness of the ΔDT curve separation, defined as:

$$\Delta DT_z = \Delta DT_{\bar{x}} \times h_{net}, \Delta DT > 0 \qquad (6)$$

where

    $\Delta DT_{\bar{x}}$ is the mean of positive ΔDT values calculated within the stratigraphic interval of interest (K1 sequence set, in this example) and

    h_{net} is the net vertical interval in feet, over which ΔDT exceeds zero within the subject interval.

$\Delta DT_{\bar{x}}$ may be used as a proxy of the overall source-rock quality in a stratigraphic interval of interest. $\Delta DT_{\bar{x}}$ and ΔDT_z were only calculated where ΔDT was greater than zero, as positive values represent higher resistivity and higher sonic travel times indicative of possible source rocks. Definition of ΔDT_z to include both the magnitude and thickness of positive ΔDT values was intended to produce an objective parameter that may be proportional to the volume of potential source rocks penetrated by each well. Following the calculation of $\Delta DT_{\bar{x}}$ and ΔDT_z for each well, the results for the lower Kingak Shale K1 sequence set were mapped (figs. 9, 10) using a gridding algorithm in IHS Kingdom® version 8.8 software. A digital workflow for calculation of $\Delta DT_{\bar{x}}$ and ΔDT_z using IHS Kingdom® 8.8 software is presented in appendix 1.

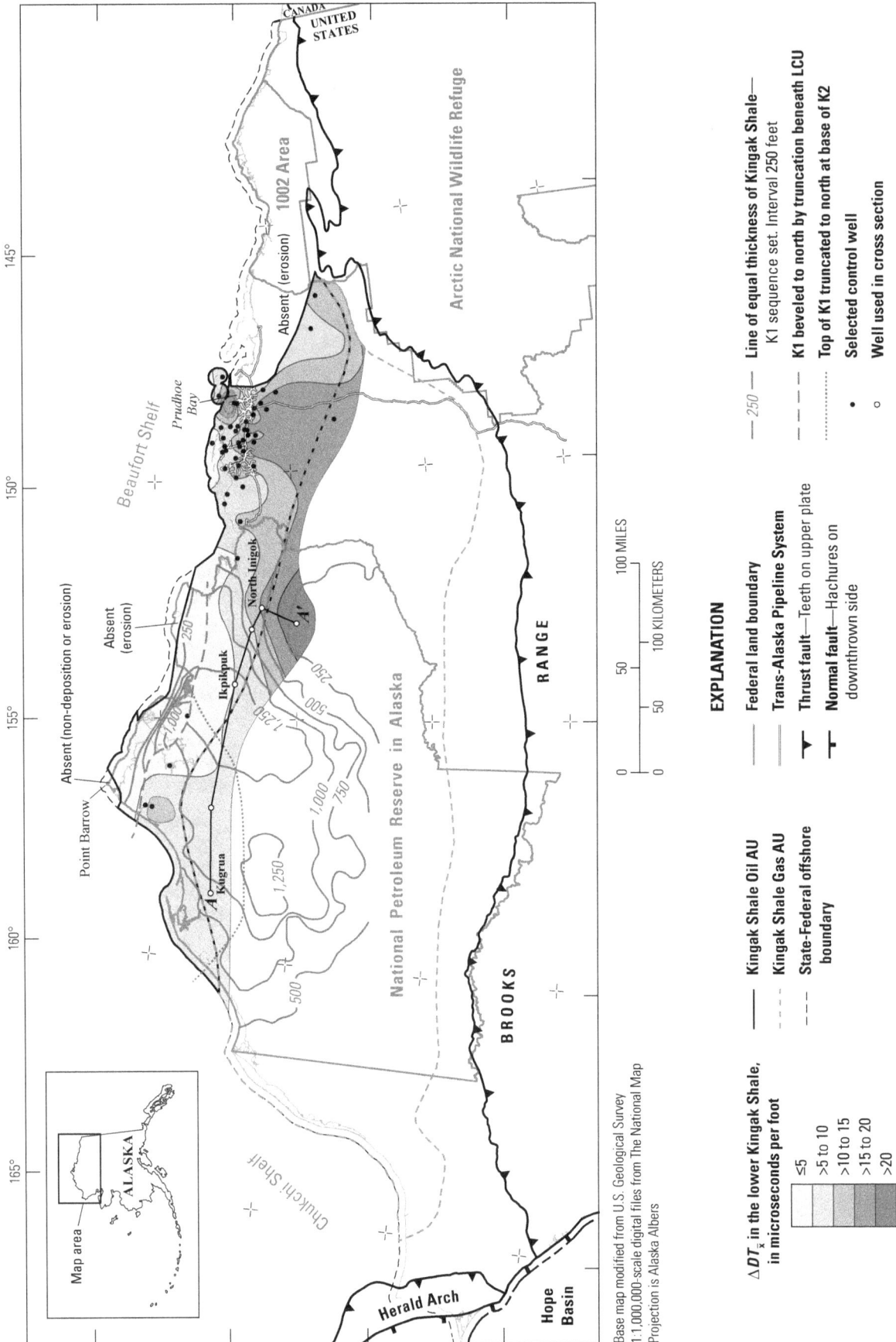

Base map modified from U.S. Geological Survey
1:1,000,000-scale digital files from The National Map
Projection is Alaska Albers

EXPLANATION

$\Delta DT_{\bar{x}}$ **in the lower Kingak Shale,**
in microseconds per foot

	≤5
	>5 to 10
	>10 to 15
	>15 to 20
	>20

— Kingak Shale Oil AU

---- Kingak Shale Gas AU

– – – State-Federal offshore
boundary

—— Federal land boundary

—— Trans-Alaska Pipeline System

▼ Thrust fault—Teeth on upper plate

┳ Normal fault—Hachures on
downthrown side

—250— Line of equal thickness of Kingak Shale—
K1 sequence set. Interval 250 feet

– – – K1 beveled to north by truncation beneath LCU

········· Top of K1 truncated to north at base of K2

● Selected control well

○ Well used in cross section

Figure 9.   Map of $\Delta DT_{\bar{x}}$ in the lower Kingak Shale for that part of the Alaska North Slope where requisite wireline-log data are available. Blue contours modified from Houseknecht and Bird, 2004. Abbreviations used: AU, assessment unit; LCU, Lower Cretaceous unconformity; $\Delta DT_{\bar{x}}$ is the mean of positive ΔDT values calculated within the stratigraphic interval of interest, and may be used as a proxy of the overall source-rock quality.

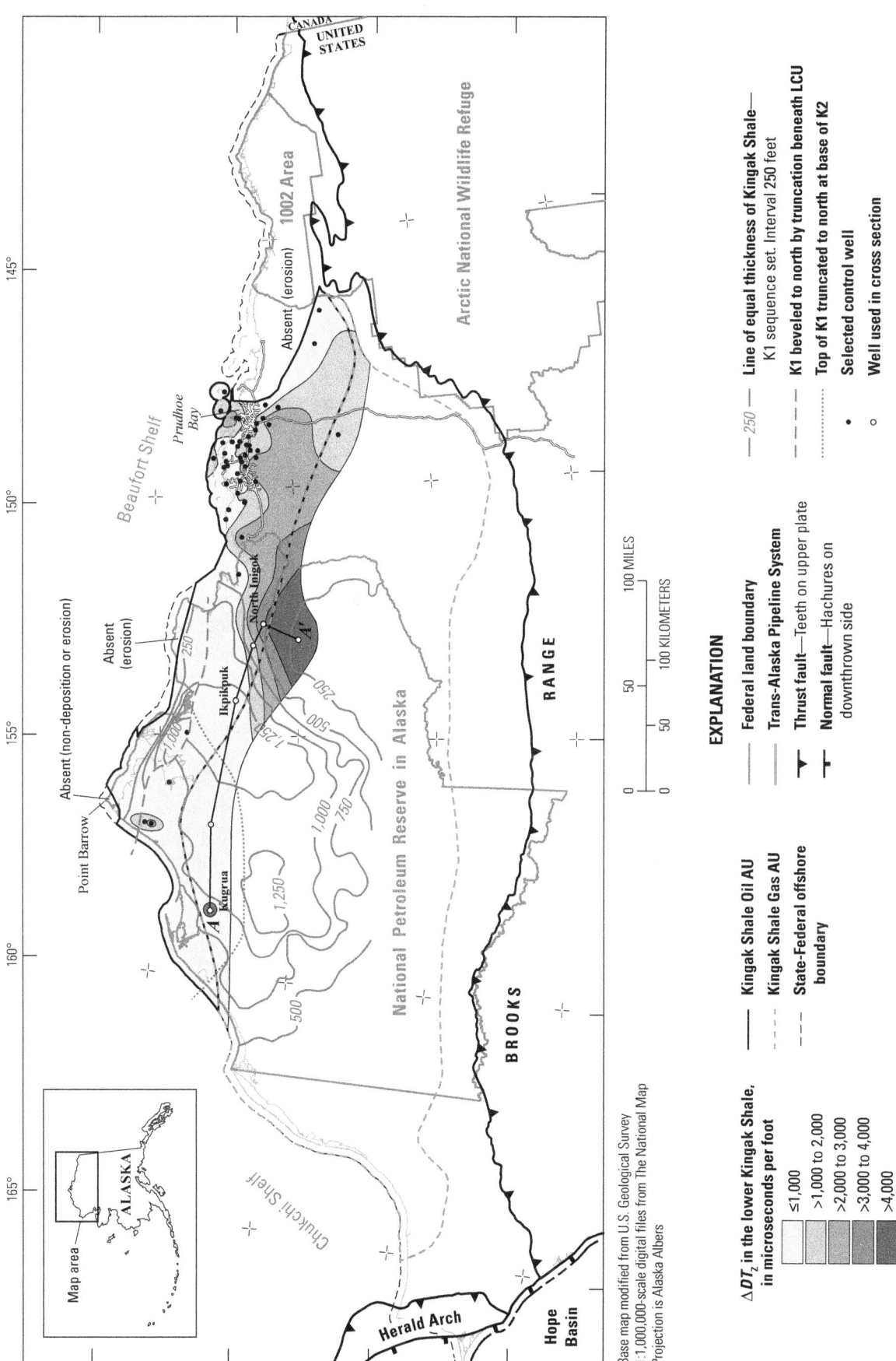

Base map modified from U.S. Geological Survey
1:1,000,000-scale digital files from The National Map
Projection is Alaska Albers

EXPLANATION

ΔDT_z **in the lower Kingak Shale,
in microseconds per foot**

≤1,000
>1,000 to 2,000
>2,000 to 3,000
>3,000 to 4,000
>4,000

— Kingak Shale Oil AU
--- Kingak Shale Gas AU
--- State-Federal offshore
      boundary

— Federal land boundary
— Trans-Alaska Pipeline System
▼ Thrust fault—Teeth on upper plate
┬ Normal fault—Hachures on
      downthrown side

—250— Line of equal thickness of Kingak Shale—
            K1 sequence set. Interval 250 feet
--- K1 beveled to north by truncation beneath LCU
········ Top of K1 truncated to north at base of K2

•   Selected control well
○   Well used in cross section

Figure 10. Map of ΔDT_z in the lower Kingak Shale for that part of the Alaska North Slope where requisite wireline-log data are available. Blue contours modified from Houseknecht and Bird, 2004. Abbreviations used: AU, assessment unit; LCU, Lower Cretaceous unconformity; ΔDT_z incorporates both the magnitude and thickness of the ΔDT curve separation, and is intended to produce an objective parameter that may be proportional to the volume of potential source rocks penetrated by each well.

Discussion

The $\Delta DT_{\bar{x}}$ and ΔDT_z parameters were developed to serve as proxies for potential source-rock quality and volume, respectively, in the absence of geochemical and thermal maturity data necessary for the direct correlation of TOC log and geochemical data. Geochemical data from the lower Kingak Shale are sparse and largely derived from cuttings collected over intervals of 10 to 100 ft (3 to 30 m). Moreover, most of the available TOC data were concentrated in relatively proximal parts of the Kingak Shale depositional system that lack organic-matter-rich and oil-prone source rocks, making a direct comparison of TOC and ΔDT in potential source-rock intervals difficult. However, where TOC measurements were available in potential source-rock intervals, ΔDT and TOC values were positively correlated (for example, North Inigok; fig. 11).

Maps of $\Delta DT_{\bar{x}}$ and ΔDT_z in the K1 sequence set reveal an increase in potential source-rock quality and volume, respectively, from north to south. The potentially richest area in the eastern NPRA corresponds to a re-entrant in the K1 shelf margin, as defined by a K1 isopach map (figs. 9, 10). Within the NPRA, $\Delta DT_{\bar{x}}$ and ΔDT_z values are inversely related to the thickness of the K1 sequence set, with the greatest values where the K1 sequence set thins distally (fig. 12). These findings agree with Houseknecht and Bird's (2004) interpretation of a distal increase in organic matter content within the K1 sequence set in the NPRA. Outside of the NPRA, $\Delta DT_{\bar{x}}$ and ΔDT_z values gradually decrease eastward toward the Arctic National Wildlife Refuge (figs. 9, 10). An exception to this regional trend occurs in the Prudhoe Bay area, where a pod of large $\Delta DT_{\bar{x}}$ and ΔDT_z values highlight a potential rich source-rock area, although the controls are not understood (figs. 9, 10).

The large ΔDT_z value calculated for the Kugrua well (location shown in figure 10) is attributed to elevated methane concentrations within shale of the K2 interval (Hayba and others, 2002), skewing the shale baseline values meant to be derived in a non-source-rock interval. This results in an offset of the DT and DT_{logR} log curves within the K1 interval, where the curves otherwise would be superimposed. This small curve separation (as evident in figure 9), combined with the thickness of the K1 sequence set preserved in the proximal portion of the basin, result in a large ΔDT_z value that is unlikely to correspond to a large volume of potential source rock.

Conclusions

The methodology outlined in this report can be used in a completely digital workflow to evaluate the richness and volume of potential source rocks, both in individual wells and in a map area containing multiple wells, provided that a non-source-rock interval of mudstone or shale can be identified to establish a baseline for comparison. Use of $\Delta DT_{\bar{x}}$ and ΔDT_z parameters delineates regional source-rock potential in the lower Kingak Shale, and map results are consistent with known patterns of lithofacies and geochemistry. However, because thermal maturity was not considered in the calculation of these parameters in the Kingak Shale test case, TOC values cannot be estimated for individual wells or regionally. Therefore, the $\Delta DT_{\bar{x}}$ and ΔDT_z mapping parameters should be viewed as first-step reconnaissance tools for identifying possible source-rock potential.

Acknowledgments

We thank Tom Wild, President and Owner of Tom Wild Petrophysical Services LLC, for insightful discussion on geophysical-log interpretation.

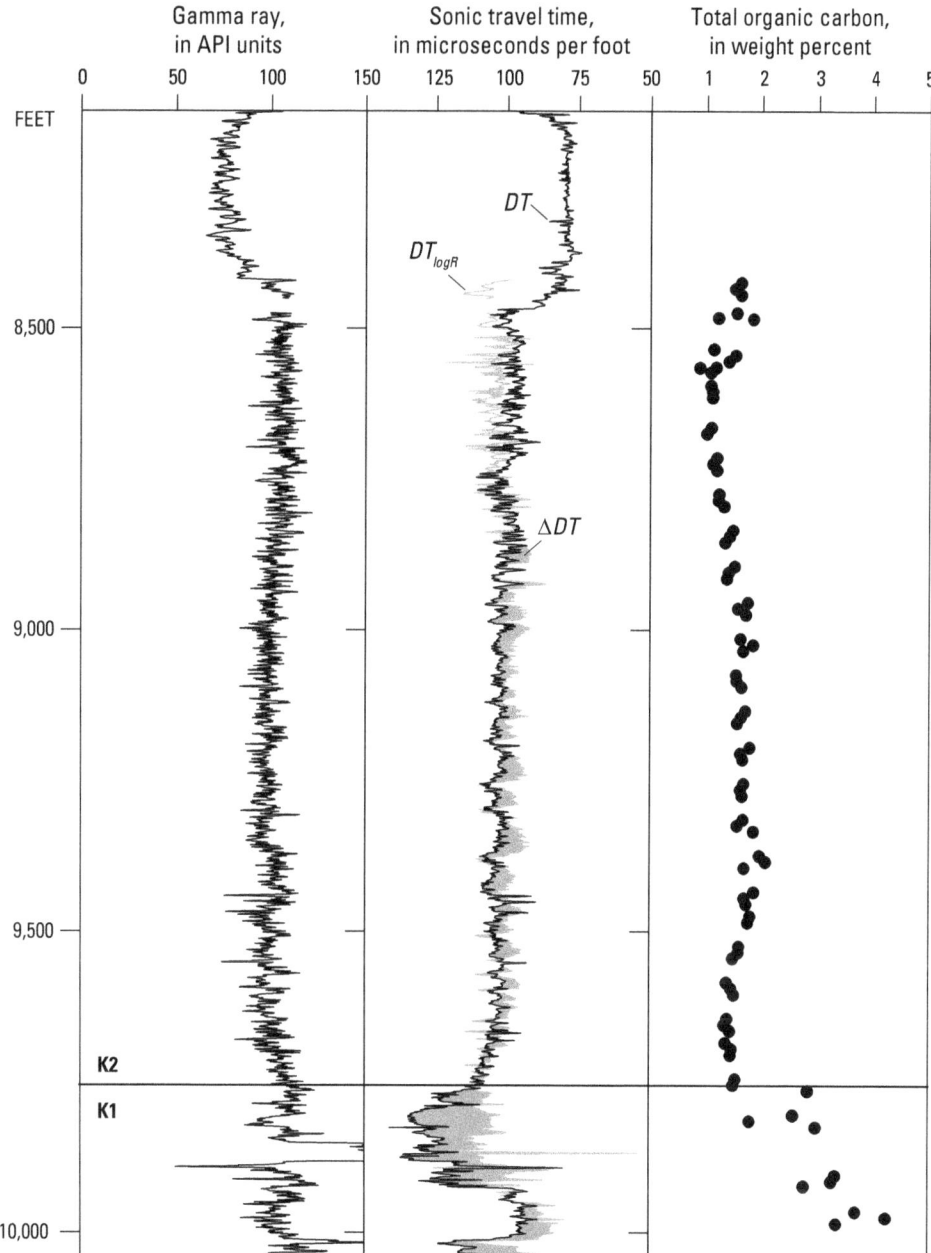

Figure 11. Gamma-ray log, sonic travel time, and total organic carbon content of the Kingak Shale K1 and K2 sequence sets in the North Inigok well (location shown in figures 9 and 10). Greater positive separation of the *DT* and *DT*$_{logR}$ curves (Δ*DT*; highlighted in red) in the K1 sequence set correlates with high total organic carbon values. Depths are below kelly bushing. Abbreviations: API, American Petroleum Institute; *DT*, sonic travel time.

A

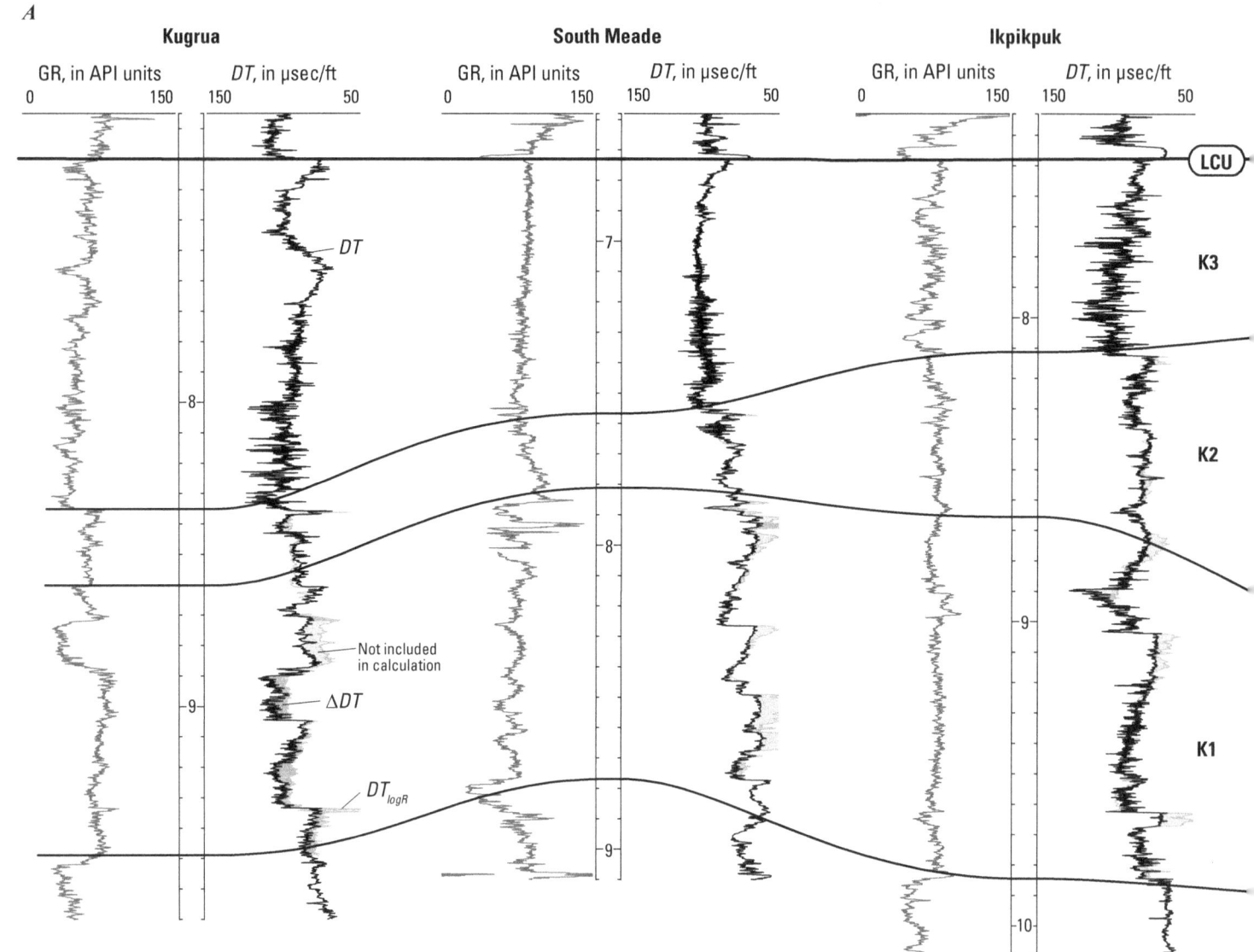

Figure 12. Well logs from cross section *A–A'* illustrating regional change in Δ*DT* (highlighted in red) related to location in Kingak Shale K1–K3 sequence sets. Intervals with curve separation highlighted in gray have less than 60 percent shale content and were not included in the calculation of Δ*DT*. Wireline-log measured depth ticks below kelly bushing are at 100-foot (ft) (30-meter) intervals. Cross section hung on the Lower Cretaceous unconformity (LCU). Location of cross section shown in figures 9 and 10. Abbreviations: API, American Petroleum Institute; *DT*, sonic travel time; GR, gamma ray; μsec/ft, microseconds per foot.

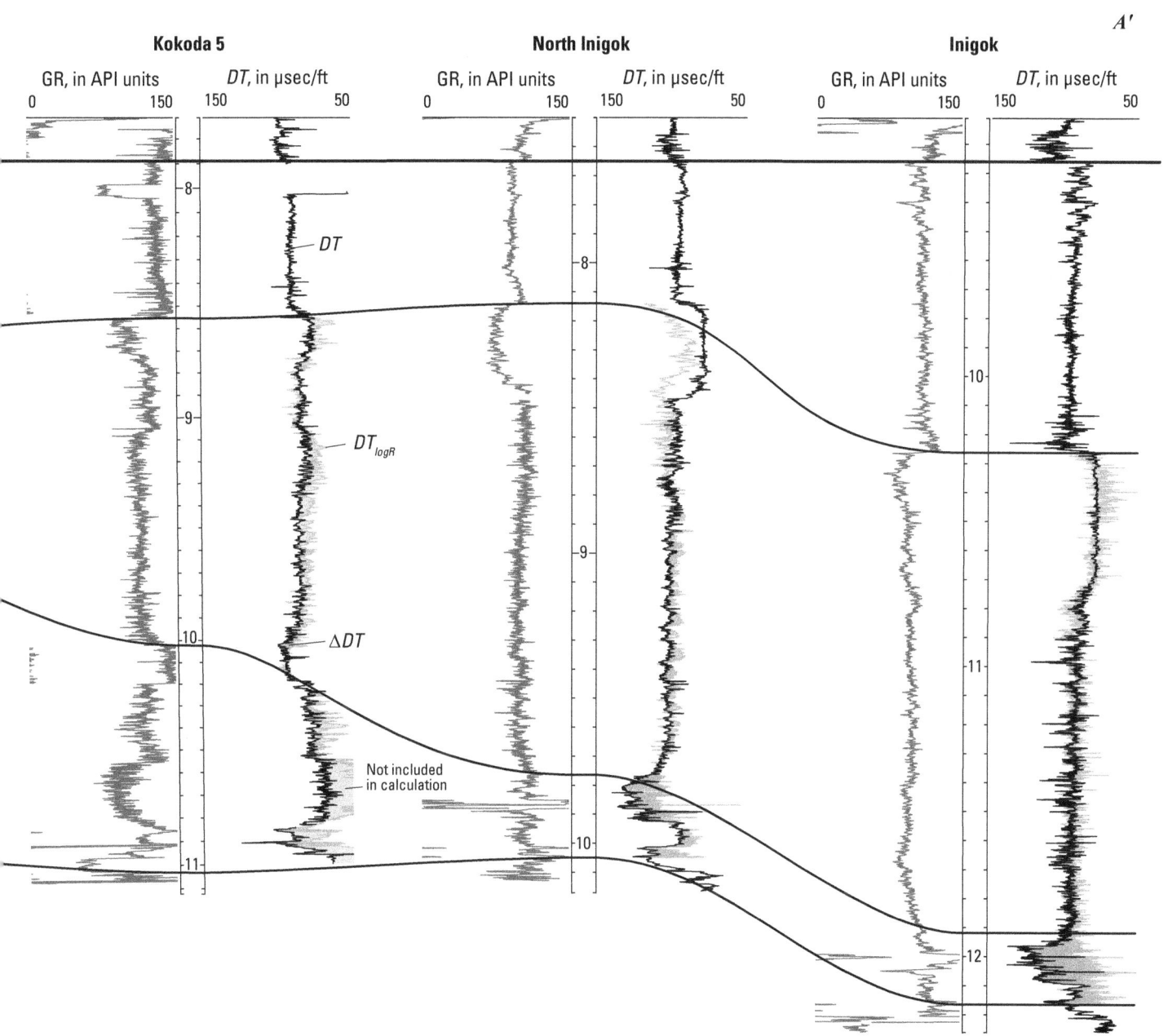

Figure 12. Well logs from cross section *A–A'* illustrating regional change in Δ*DT* (highlighted in red) related to location in Kingak Shale K1–K3 sequence sets. Intervals with curve separation highlighted in gray have less than 60 percent shale content and were not included in the calculation of Δ*DT*. Wireline-log measured depth ticks below kelly bushing are at 100-foot (ft) (30-meter) intervals. Cross section hung on the Lower Cretaceous unconformity (LCU). Location of cross section shown in figures 9 and 10. Abbreviations: API, American Petroleum Institute; *DT*, sonic travel time; GR, gamma ray; μsec/ft, microseconds per foot.—Continued

References Cited

Beers, R.F., 1945, Radioactivity and organic content of some Paleozoic shales: American Association of Petroleum Geologists Bulletin, v. 29, no. 1, p. 1–22. [Also available at http://archives.datapages.com/data/bulletns/1944-48/data/pg/0029/0001/0000/0001.htm.]

Bird, K.J., 1985, The framework geology of the North Slope of Alaska as related to oil-source rock correlations; Introductory papers, in Magoon, L.B., and Claypool, G.E., eds., Alaska North Slope oil-rock correlation study: American Association of Petroleum Geologists Studies in Geology 20, p. 3–29. [Also available at http://archives.datapages.com/data/specpubs/geochem1/data/a031/a031/0001/0000/0003.htm.]

Bird, K.J., 2001, Alaska; A twenty-first-century petroleum province, chap. 9 of Downey, M.W., Threet, J.C., and Morgan, W.A., eds., Petroleum provinces of the twenty-first century: American Association of Petroleum Geologists Memoir 74, p. 137–165. [Also available at http://archives.datapages.com/data/specpubs/memoir74/m74ch09/m74ch09.htm.]

Bird, K.J., and Houseknecht, D.W., 2011, Geology and petroleum potential of the Arctic Alaska petroleum province, chap. 32 of Spencer, A.M., Embry, A.F., Gautier, D.L., Stoupakova, A.V., and Sørensen, Kai, eds., Arctic petroleum geology: Geological Society of London Memoirs, v. 35, p. 485–499. [Also available at http://dx.doi.org/10.1144/M35.32.]

Bowman, Thomas, 2010, Direct method for determining organic shale potential from porosity and resistivity logs to identify possible resource plays: American Association of Petroleum Geologists Search and Discovery Article 110128 [adapted from oral presentation at AAPG Annual Convention, New Orleans, Louisiana, April 11–14, 2010], 34 p., accessed November 24, 2014, at http://www.search-anddiscovery.com/documents/2010/110128bowman/ndx_bowman.pdf.

Carman, G.J., and Hardwick, Peter, 1983, Geology and regional setting of the Kuparuk oil field, Alaska: American Association of Petroleum Geologists Bulletin, v. 67, no. 6, p. 1014–1031. [Also available at http://archives.datapages.com/data/bulletns/1982-83/data/pg/0067/0006/1000/1014.htm.]

Claypool, G.E., and Magoon, L.B., 1985, Comparison of oil-source rock correlation data for Alaskan North Slope—Techniques, results, and conclusions, in Magoon, L.B., and Claypool, G.E., eds., Alaska North Slope oil-rock correlation study: American Association of Petroleum Geologists Studies in Geology 20, p. 49–81. [Also available at http://archives.datapages.com/data/specpubs/geochem1/data/a031/a031/0001/0000/0049.htm.]

Dellenbach, J., Espitalie, J., and Lebreton, F., 1983, Source rock logging: Transactions of the 8th European Society of Petrophysicists and Well Log Analysts Symposium, paper D.

Gradstein, F.M., and Ogg, J.G., 1996, A Phanerozoic time scale: Episodes, v. 19, no. 1–2, p. 3–5.

Gradstein, F.M., Ogg, J.G., Smith, A.G., Bleeker, W., and Lourens, L.J., 2004, A new geologic time scale, with special reference to Precambrian and Neogene: Episodes, v. 27, no. 2, p. 83–100.

Hayba, D.O., Bird, K.J., and Garrity, C.P., 2002, Subsurface oil and gas indications in and adjacent to the NPRA, chap. 5 of Houseknecht, D.W., ed., National Petroleum Reserve, Alaska (NPRA) core images and well data: U.S. Geological Survey Digital Data Series DDS–75 (4 CD–ROM).

Hood, A., Gutjahr, C.C.M., and Heacock, R.L., 1975, Organic metamorphism and the generation of petroleum: American Association of Petroleum Geologists Bulletin, v. 59, no. 6, p. 986–996. [Also available at http://archives.datapages.com/data/bulletns/1974-76/data/pg/0059/0006/0950/0986.htm.]

Houseknecht, D.W., and Bird, K.J., 2004, Sequence stratigraphy of the Kingak Shale (Jurassic–Lower Cretaceous), National Petroleum Reserve in Alaska: American Association of Petroleum Geologists Bulletin, v. 88, no. 3, p. 279–302. [Also available at http://archives.datapages.com/data/bulletns/2004/03mar/0279/0279.htm.]

Houseknecht, D.W., Bird, K.J., and Garrity, C.P., 2012, Assessment of undiscovered petroleum resources of the Arctic Alaska Petroleum Province: U.S. Geological Survey Scientific Investigations Report 2012–5147, 26 p. [Also available at http://pubs.usgs.gov/sir/2012/5147/.]

Houseknecht, D.W., Rouse, W.A., and Garrity, C.P., 2012, Arctic Alaska shale-oil and shale-gas resource potential: Arctic Technology Conference, Houston, Texas, December 3–5, 2012, Offshore Technology Conference, 4 p. [Also available at http://dx.doi.org/10.4043/23796-MS.]

Houseknecht, D.W., Rouse, W.A., Garrity, C.P., Whidden, K.J., Dumoulin, J.A., Schenk, C.J., Charpentier, R.R., Cook, T.A., Gaswirth, S.B., Kirschbaum, M.A., and Pollastro, R.M., 2012, Assessment of potential oil and gas resources in source rocks of the Alaska North Slope, 2012: U.S. Geological Survey Fact Sheet 2012–3013, 2 p., accessed November 24, 2014, at http://pubs.usgs.gov/fs/2012/3013/.

Houseknecht, D.W., Rouse, W.A., Paxton, S.T., Mars, J.C., and Fulk, Bryant, 2014, Upper Devonian–Mississippian stratigraphic framework of the Arkoma Basin and distribution of potential source-rock facies in the Woodford–Chattanooga and Fayetteville–Caney shale-gas systems: American Association of Petroleum Geologists Bulletin, v. 98, no. 9, p. 1739–1759. [Also available at http://dx.doi.org/10.1306/03031413025.]

Hubbard, R.J., Edrich, S.P., and Rattey, R.P., 1987, Geologic evolution and hydrocarbon habitat of the 'Arctic Alaska microplate': Marine and Petroleum Geology, v. 4, no. 1, p. 2–34. [Also available at http://dx.doi.org/10.1016/0264-8172(87)90019-5.]

Lerand, Monti, 1973, Beaufort Sea, *in* McCrossan, R.G., ed., The future petroleum provinces of Canada—Their geology and potential: Canadian Society of Petroleum Geologists Memoir 1, p. 315–386.

Magoon, L.B., and Claypool, G.E., 1984, The Kingak Shale of northern Alaska—Regional variations in organic geochemical properties and petroleum source rock quality: Organic Geochemistry, v. 6, p. 533–542. [Also available at http://dx.doi.org/10.1016/0146-6380(84)90076-7.]

Magoon, L.B., Lillis, P.G., Bird, K.J., Lampe, C., and Peters, K.E., 2003, Alaskan North Slope petroleum systems: U.S. Geological Survey Open-File Report 03–324, 3 sheets, accessed November 24, 2014, at http://pubs.usgs.gov/of/2003/of03-324/.

Masterson, W.D., IV, 2001, Petroleum filling history of central Alaskan North Slope fields: University of Texas at Dallas, unpublished Ph.D. dissertation, 222 p.

Masterson, W.D., and Paris, C.E., 1987, Depositional history and reservoir description of the Kuparuk River Formation, North Slope, Alaska, *in* Tailleur, Irv, and Weimer, Paul, eds., Alaskan North Slope Geology: Pacific Section, Society of Economic Paleontologists and Mineralogists and Alaska Geological Society 50, v. 1, p. 95–107.

Meissner, F.F., 1978, Petroleum geology of the Bakken Formation, Williston basin, North Dakota and Montana, *in* The economic geology of the Williston basin: Montana Geological Society, 24th annual Williston Basin Symposium, Billings, Montana, September 24–27, 1978, p. 207–227.

Meyer, B.L., and Nederlof, M.H., 1984, Identification of source rocks on wireline logs by density/resistivity and sonic transit time/resistivity cross plots: American Association of Petroleum Geologists Bulletin, v. 68, no. 2, p. 121–129. [Also available at http://archives.datapages.com/data/bulletns/1984-85/data/pg/0068/0002/0100/0121.htm.]

Mull, C.G., Houseknecht, D.W., and Bird, K.J., 2003, Revised Cretaceous and Tertiary stratigraphic nomenclature in the Colville Basin, northern Alaska: U.S. Geological Survey Professional Paper 1673, 51 p., accessed November 24, 2014, at http://pubs.usgs.gov/pp/p1673/p1673.pdf.

Nixon, R.P., 1973, Oil source beds in Cretaceous Mowry Shale of northwestern interior United States: American Association of Petroleum Geologists Bulletin, v. 57, no. 1, p. 136–161. [Also available at http://archives.datapages.com/data/bulletns/1971-73/data/pg/0057/0001/0100/0136.htm.]

Passey, Q.R., Bohacs, K.M., Esch, W.L., Klimentidis, R., and Sinha, S., 2010, From oil-prone source rock to gas-producing shale reservoir—Geologic and petrophysical characterization of unconventional shale-gas reservoirs: International Oil and Gas Conference and Exhibition, Beijing, China, June 8–10, 2010, Society of Petroleum Engineers Paper 131350, 29 p.

Passey, Q.R., Creaney, S., Kulla, J.B., Moretti, F.J., and Stroud, J.D., 1990, A practical model for organic richness from porosity and resistivity logs: American Association of Petroleum Geologists Bulletin, v. 74, no. 12, p. 1777–1794. [Also available at http://archives.datapages.com/data/bulletns/1990-91/data/pg/0074/0012/0000/1777.htm.]

Peters, K.E., Magoon, L.B., Bird, K.J., Valin, Z.C., and Keller, M.A., 2006, North Slope, Alaska—Source-rock distribution, richness, thermal maturity, and petroleum charge: American Association of Petroleum Geologists Bulletin, v. 90, no. 2, p. 261–292. [Also available at http://dx.doi.org/10.1306/09210505095.]

Premuzic, E.T., Gaffney, J.S., and Manowitz, B., 1986, The importance of sulfur isotope ratios in the differentiation of Prudhoe Bay crude oils: Journal of Geochemical Exploration, v. 26, no. 2, p. 151–159. [Also available at http://dx.doi.org/10.1016/0375-6742(86)90064-6.]

Schmoker, J.W., 1979, Determination of organic content of Appalachian Devonian shales from formation-density logs: American Association of Petroleum Geologists Bulletin, v. 63, no. 9, p. 1504–1509. [Also available at http://archives.datapages.com/data/bulletns/1977-79/data/pg/0063/0009/1500/1504.htm.]

Schmoker, J.W., 1981, Determination of organic-matter content of Appalachian Devonian shales from gamma-ray logs: American Association of Petroleum Geologists Bulletin, v. 65, no. 7, p. 1285–1298. [Also available at http://archives.datapages.com/data/bulletns/1980-81/data/pg/0065/0007/1250/1285.htm.]

Schmoker, J.W., and Hester, T.C., 1989, Oil generation inferred from formation resistivity—Bakken Formation, Williston basin, North Dakota: Transactions of the Thirtieth Society of Petrophysicists and Well Log Analysts Annual Logging Symposium, paper H.

Sedivy, R.A., Penfield, I.E., Halpern, H.I., Drozd, R.J., Cole, G.A., and Burwood, R., 1987, Investigation of source rock-crude oil relationships in the northern Alaska hydrocarbon habitat, *in* Tailleur, Irv, and Weimer, Paul, eds., Alaskan North Slope Geology: Pacific Section, Society of Economic Paleontologists and Mineralogists and Alaska Geological Society 50, v. 1, p. 169–179.

Seifert, W.K., Moldowan, J.M., and Jones, R.W., 1980, Application of biological marker chemistry to petroleum exploration: Proceedings of the 10th World Petroleum Congress, Bucharest, Romania, September 9–14, 1979, p. 425–440.

Sondergeld, C.H., Newsham, K.E., Comisky, J.T., Rice, M.C., and Rai, C.S., 2010, Petrophysical considerations in evaluating and producing shale gas resources: Society of Petroleum Engineers Unconventional Gas Conference, Pittsburgh, Pennsylvania, February 23–25, 2010, Society of Petroleum Engineers Paper 131768, 34 p.

Threlkeld, C.N., Obuch, R.C., and Gunther, G.L., 2000, Organic geochemistry data of Alaska: U.S. Geological Survey Digital Data Series DDS–59, accessed November 24, 2014, at http://pubs.usgs.gov/dds/dds-059/.

Tissot, B.P., and Welte, D.H., 1984, Petroleum formation and occurrence: New York, Springer-Verlag, 699 p. [Also available at http://dx.doi.org/10.1007/978-3-642-96446-6.]

Appendix 1. Workflow for Calculating Key Parameters

The following workflow was compiled using IHS Kingdom® version 8.8 software with the EarthPAK® module. This workflow assumes that a project has already been populated with well locations, digital gamma-ray, resistivity, and sonic wireline-log data in LAS format, and formation tops that bound the stratigraphic interval of interest. For instructions on how to input these data, the reader is referred to the IHS Kingdom® version 8.8 software help files. This workflow uses input parameters specific to the Alaska North Slope Kingak Shale that may not be applicable to all shale plays.

Calculating Shale Volume (V_{sh})

Note: The Petrophysics module is available with an EarthPAK license.

1. From the **Logs** menu, select **Petrophysics...** to proceed to the **Petrophysics** dialog box.

2. In the **Quick-Look Analysis Navigator** window, select **Zone-related Parameters**.

3. In the **Zone-related Parameters** window, from the **Select an Existing Zone** drop-down menu, select **Borehole (Public)**. Under **Select Lithology and Porosity Model**, from the **Reservoir Lithology** drop-down menu, select the appropriate reservoir lithology (sandstone in this example). Accept all other defaults.

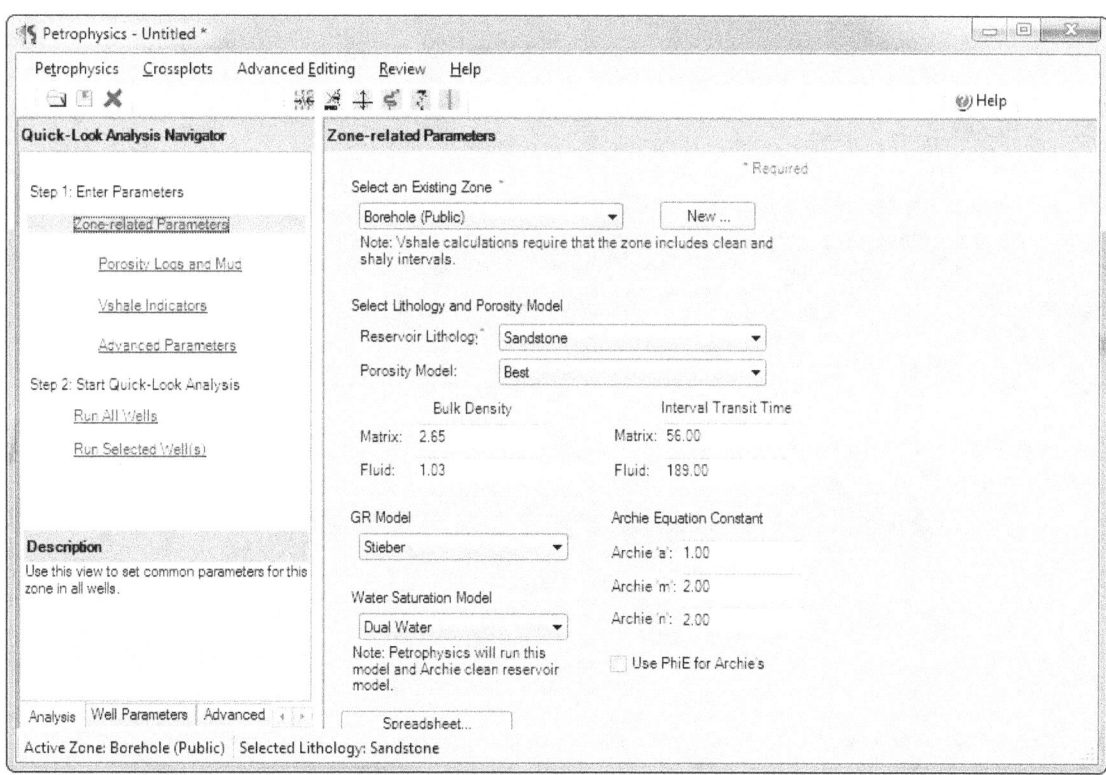

4. In the **Quick-Look Analysis Navigator**, under **Zone-related Parameters**, select **Vshale Indicators**. Calculated values for clean and shale baselines for gamma-ray (GR) and spontaneous potential (SP) are displayed; these may be accepted or manually adjusted.

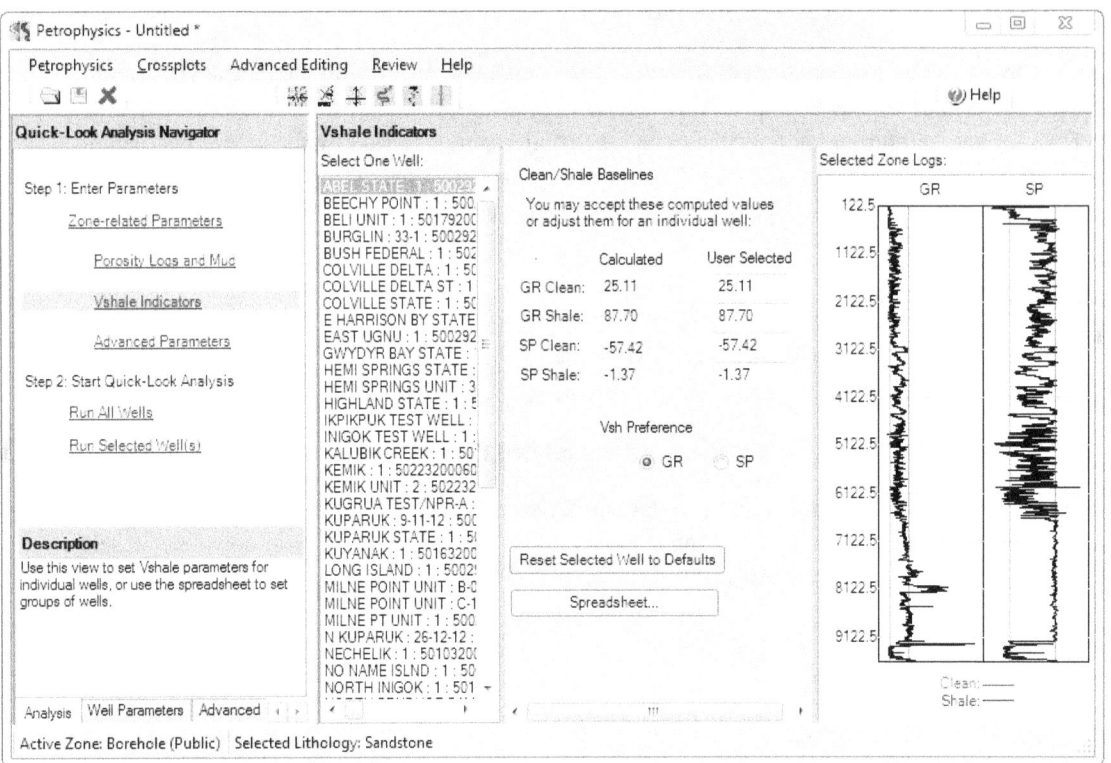

5. In the **Quick-Look Analysis Navigator**, select **Run All Wells** to proceed to the **Select Logs to Save** dialog box.

6. Select **VShale** from the curve selection window.

7. Under **Save Options**, toggle **Create New Curves Only** and accept the default prefix of **SMT**.

8. Click **OK**.

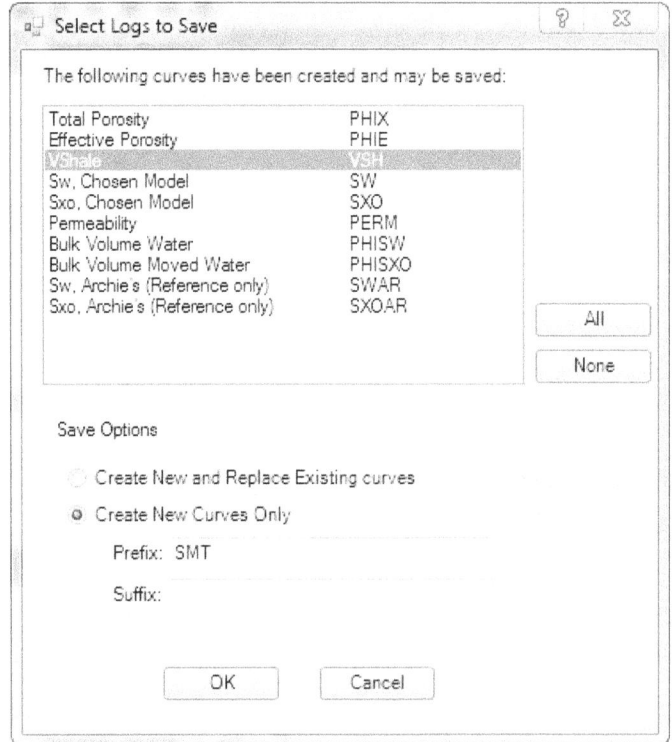

Calculating *Log R*

1. From the **Logs** menu, select **Calculations** → **Equation** to proceed to the **Select or Enter an Equation** dialog box.

2. In the **Equation Category** window, type **Other**.

3. In the **Equation** window, type **LOGR=**. In the **Functions** window below, double click on **LOG10(x)** to place it in the equation and replace (x) with **R**. The resulting equation should read **LOGR=LOG10(R).**

4. In the **Description** window, type **LogR from resistivity**.

5. Click **Next >** to proceed to the **Assign Variables** dialog box.

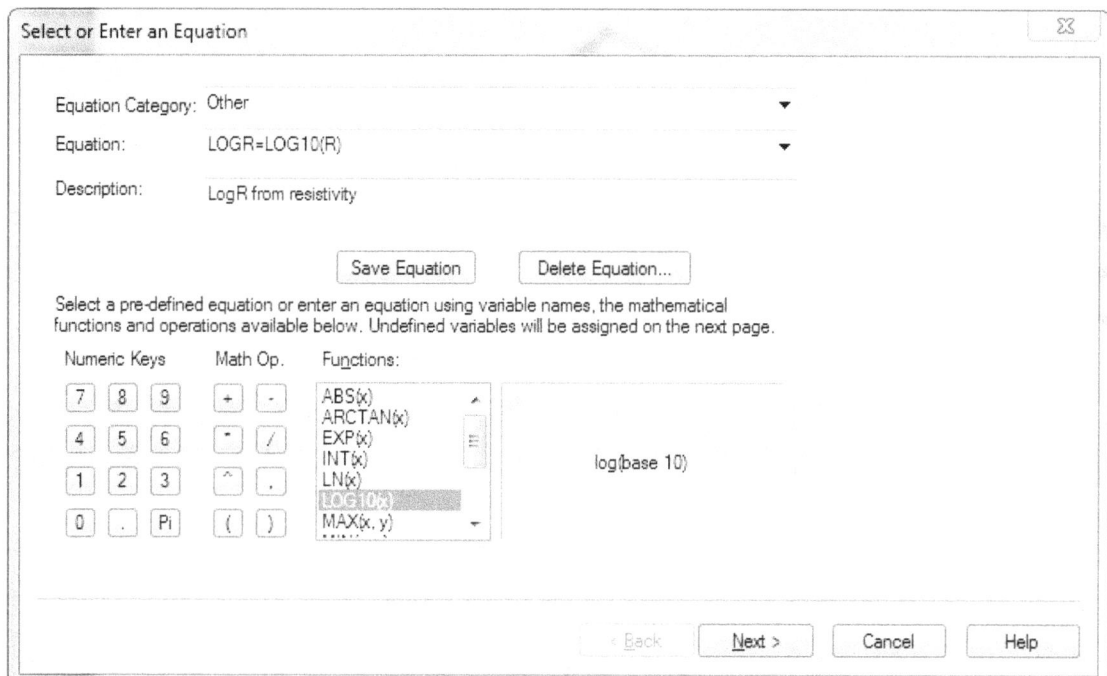

6. To assign **R** to the resistivity **(RILD)** log curve, in the **Select the Variable to Assign** window, select **R**; then toggle on **Log Curves**, and select **RILD** from the drop-down menu and click **Assign**.

7. Click **Next >** to proceed to the **Depth Range Selection** dialog box.

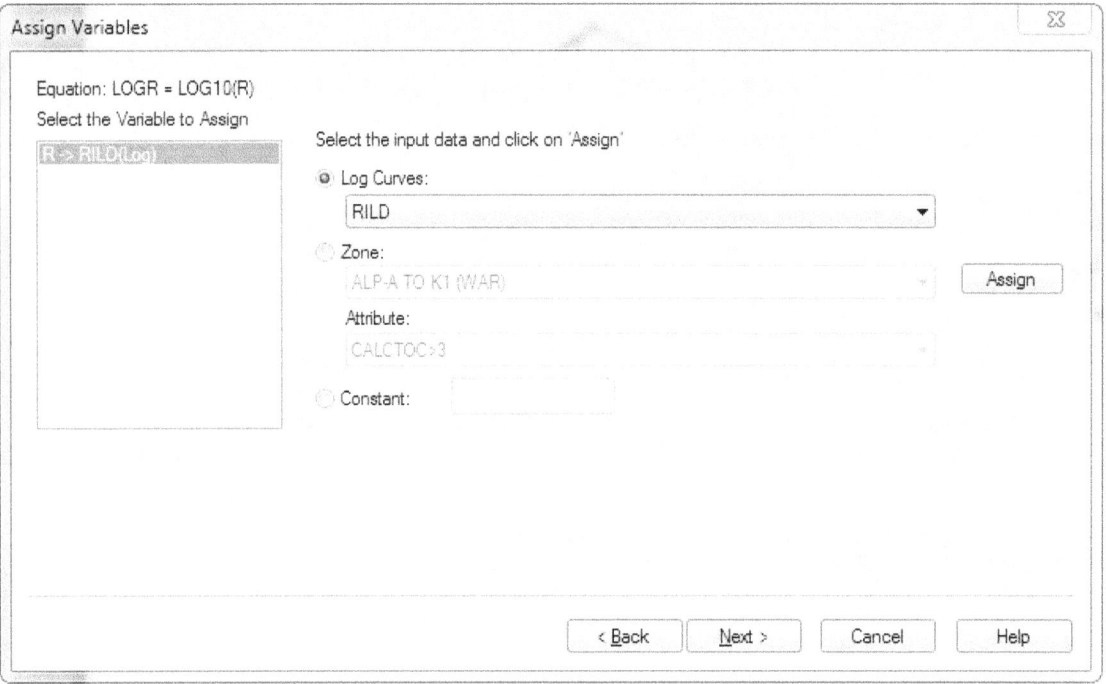

8. Toggle on **by Zone Intervals** and select **Borehole (Public)** from the drop-down menu.

9. Click **Next >** to proceed to the **Output** dialog box.

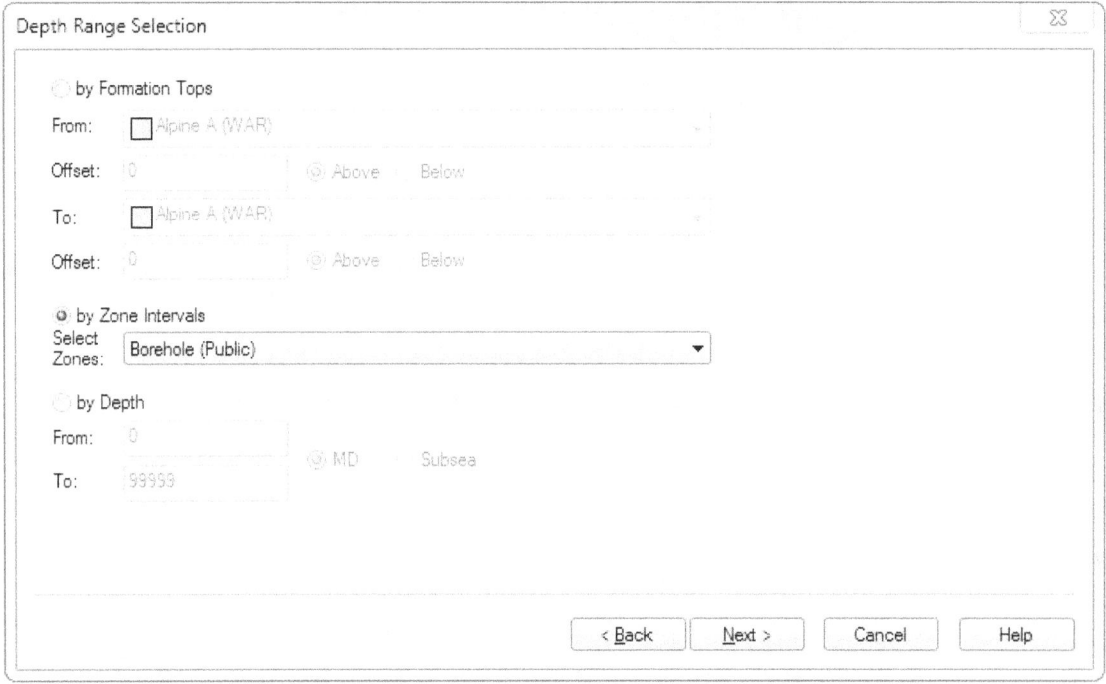

10. In the **Output Log Curve Name** window, type **LOGR**.

11. From the **Output Log Curve Type** drop-down menu, select **Other**.

12. Toggle on **Create new log curve and replace existing curve**.

13. Click **Finish**.

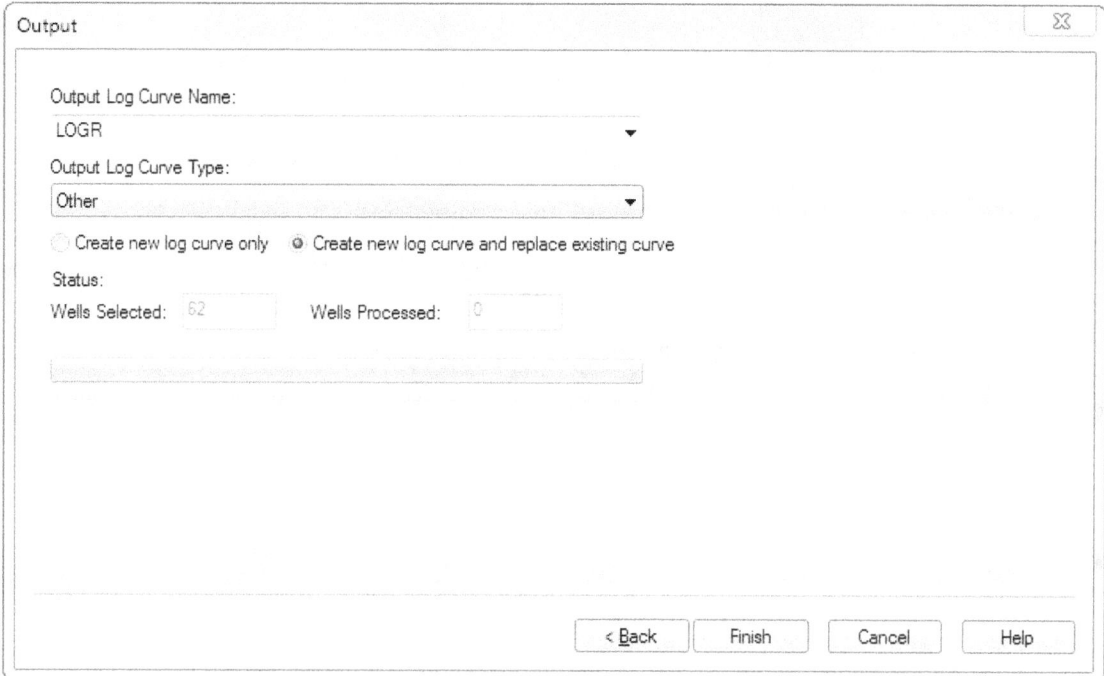

Creating a Cross Plot of *Log R* versus *DT*

Note: The following exercise assumes that the user has predefined zones within the interval of interest. For these exercises, the **K1 TO KBASE** zone is the interval that is assessed for probable source rock potential and the **K2 TO K1** zone is the assumed non-source rock interval used to create baseline sonic and resistivity data.

1. From the **Tools** menu, select **Crossplot → New** to proceed to the **Select Data** dialog box.

2. In the **Category** window, select **Log**.

3. In the **Attributes** drop-down menu, select a well that contains gamma-ray, sonic, and resistivity log curves and for which you have calculated **VSH** and **LOGR**.

4. In the **Attributes** window, select the **LOGR** curve; from the **Axis/Filter** window, select **X** and click **>**.

5. In the **Attributes** window, select the **DT** curve; from the **Axis/Filter** window, select **Y** and click **>**.

6. In the **Attributes** window, select the **SMTVSH** curve; from the **Axis/Filter** window, select **Filter** and click **>**. In the **Min Value** window, type **0.6** and keep the default **Max Value**.

7. Click **Depth Range** to proceed to the **Crossplot Depth Range Selection** dialog box.

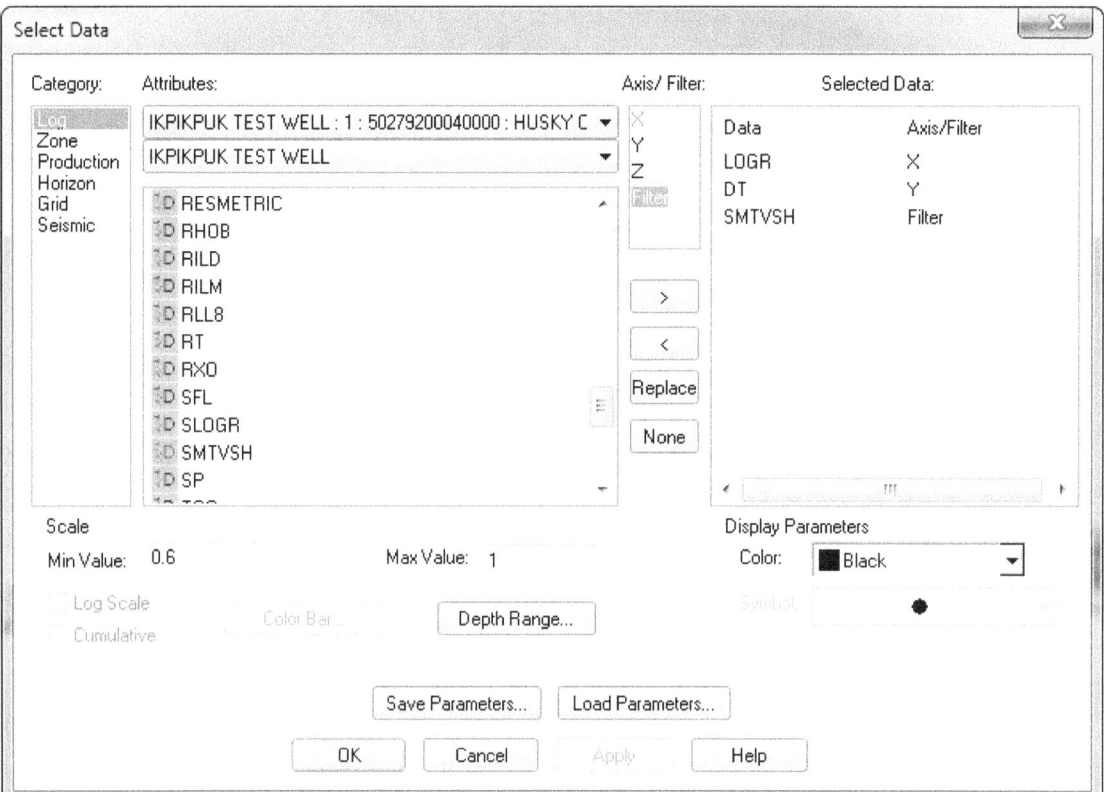

8. Toggle on **By Zone Intervals** and select the assumed non-source rock interval zone.

9. Click **OK** to return to proceed to the **Select Data** dialog box.

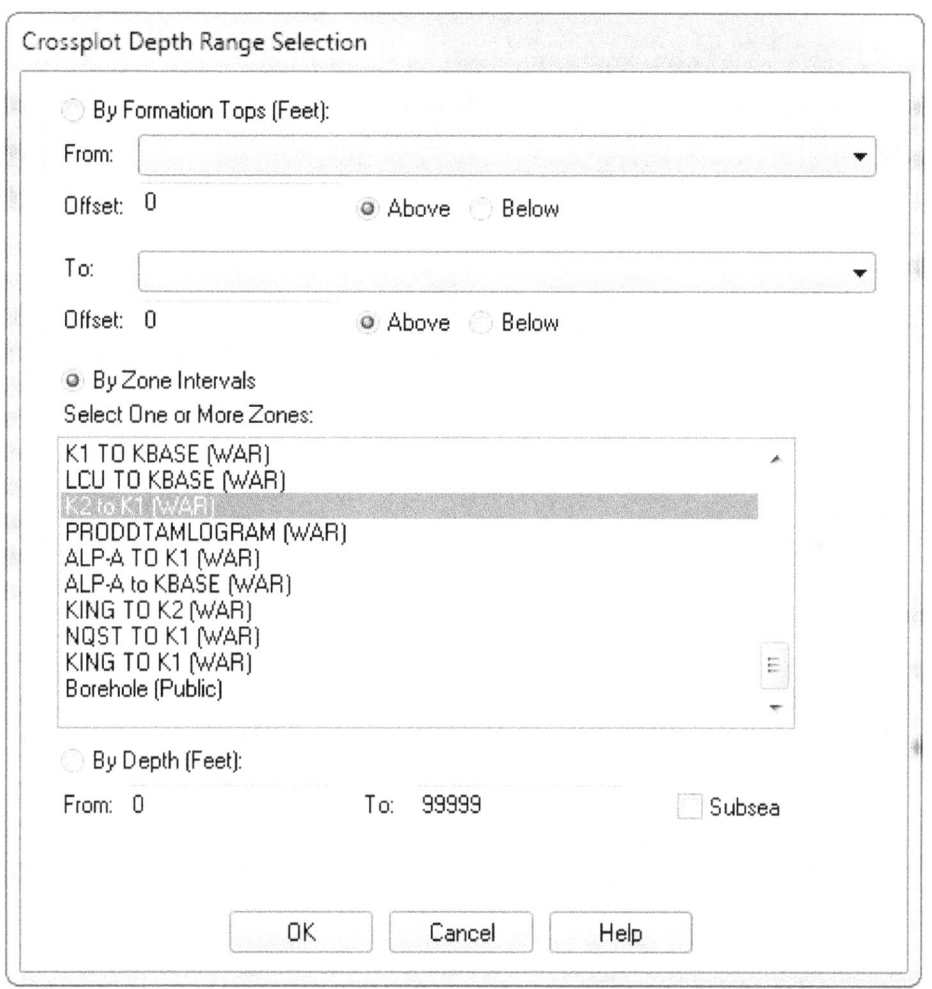

10. Click **OK** to proceed to the **Crossplot** window.

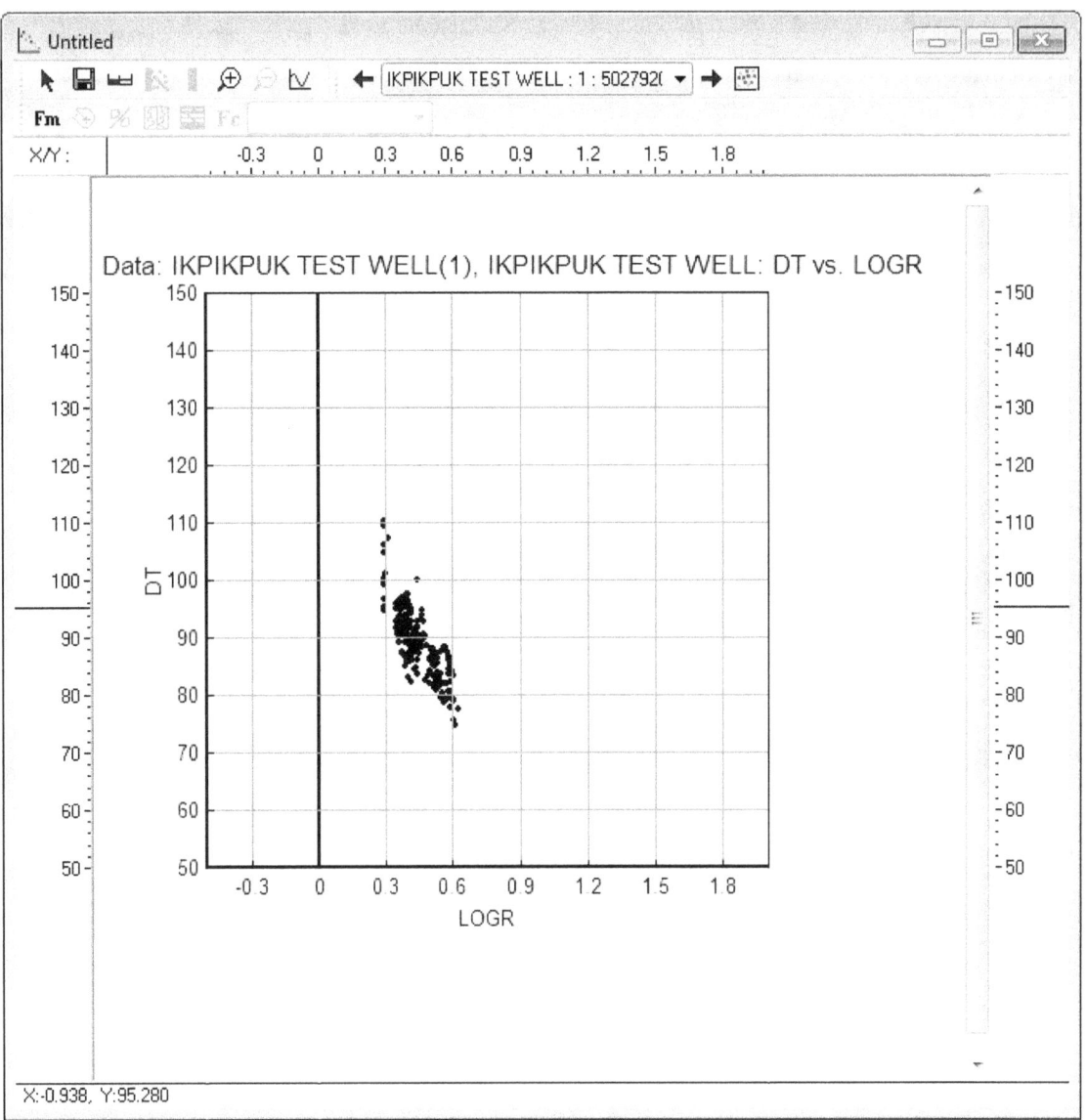

11.   Click on the regression icon 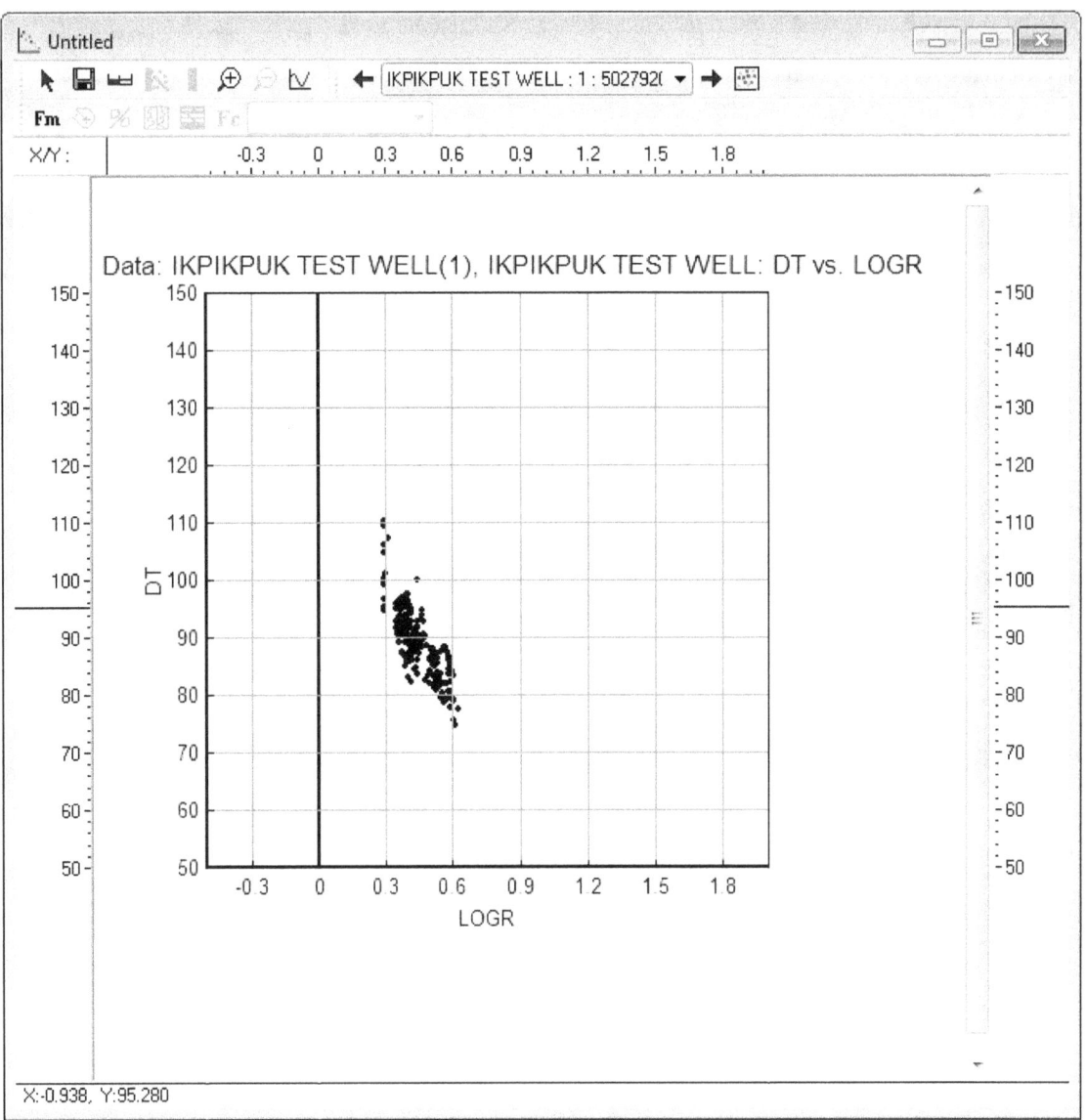 to proceed to the **Regression** dialog box.

12. Under **Regression Method**, toggle on **Numerical Regression**.

13. Under **Equation**, toggle on **Polynomial** and **Reduced Major Axis (RMA)**.

14. Click **OK** to return to the **Crossplot** window.

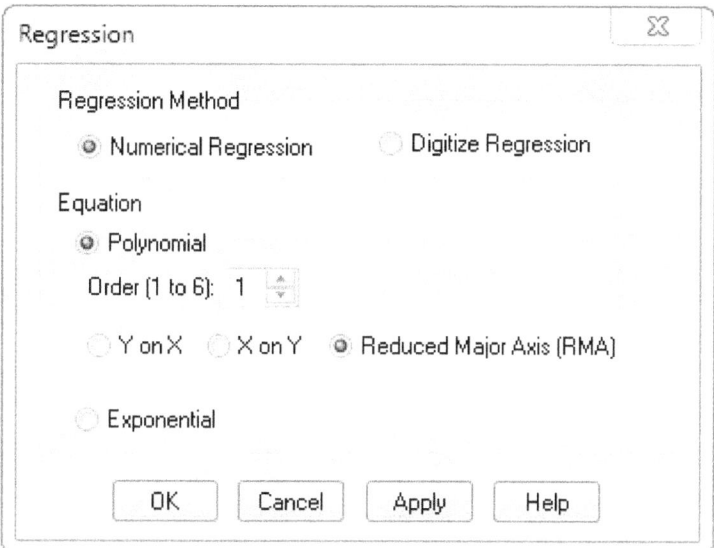

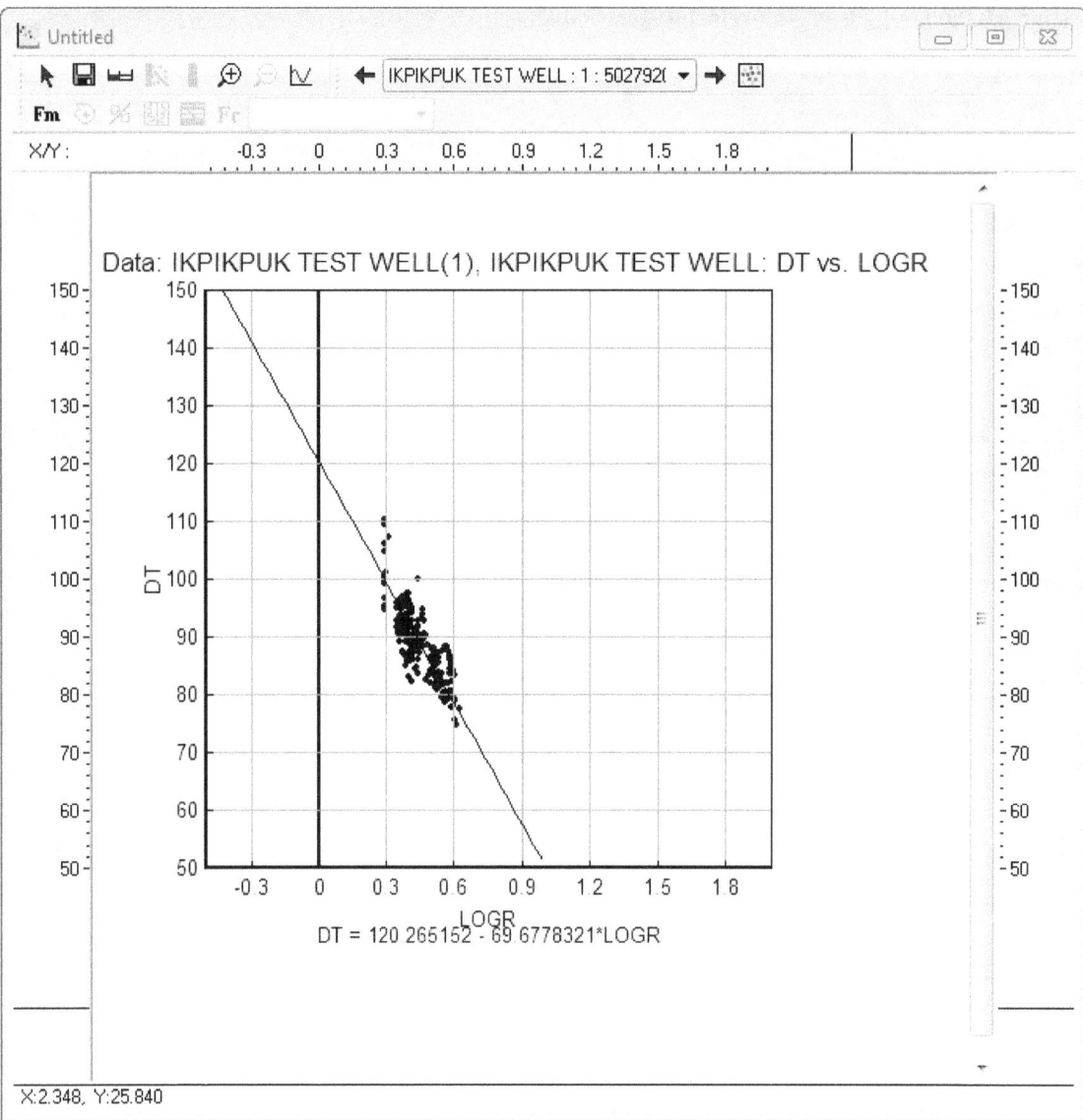

15. The **regression equation** is shown under the x-axis. Record the **slope** and **intercept** of the regression line for each well in order to calculate **DTLOGR**.

Calculating DT_{logR} (DTLOGR)

Note: Calculation of DT_{logR} must be done on a well-by-well basis. When using this workflow, make sure that only one well is selected in the project, otherwise the calculation will be applied to all wells selected.

1. From the Logs menu, select **Calculations → Equation** to proceed to the **Select or Enter an Equation** dialog box.

2. In the **Equation Category** window, type **Other**.

3. In the **Equation** window, type **DTLOGR=B-M*LOGR**.

4. In the **Description** window, type **DTLOGR from Crossplot of LogR vs. DT**.

5. Click **Next >** to proceed to the **Assign Variables** dialog box.

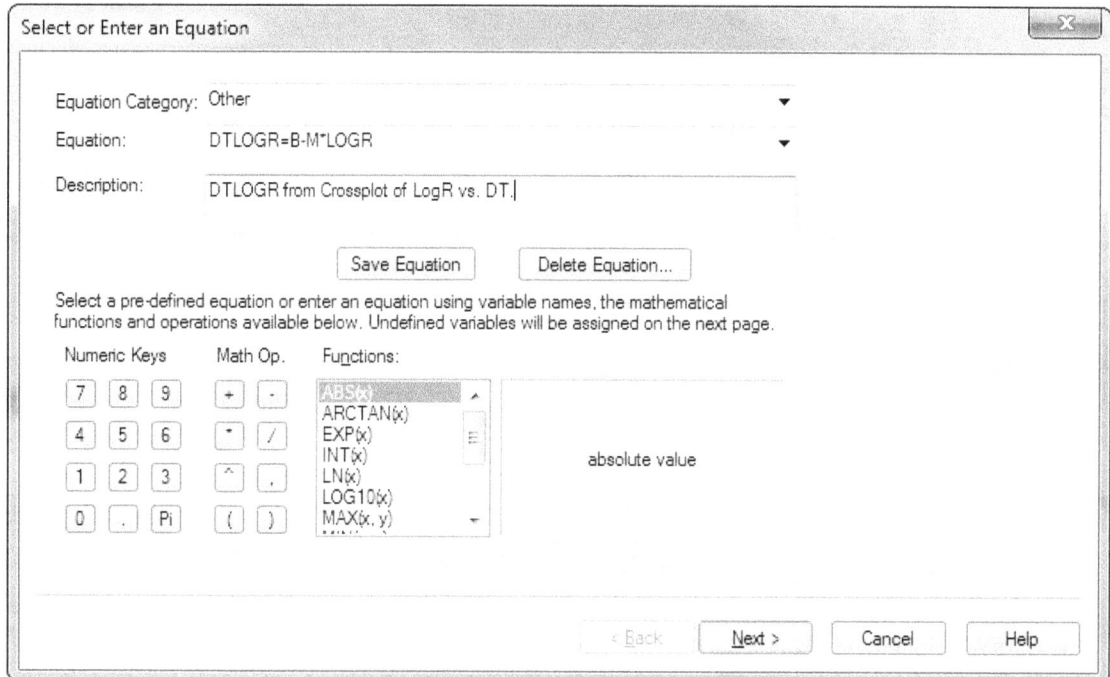

6. Assign **B** to the **y intercept** of the regression equation for the well by selecting **B** in the **Select the Variable to Assign** window, toggling on **Constant**, entering the **intercept** value in the window, and clicking **Assign**.

7. Assign **LOGR** to the **LOGR** log curve by selecting **LOGR** in the **Select the Variable to Assign** window, toggling on **Log Curves**, selecting **LOGR** from the drop-down menu, and clicking **Assign**.

8. Assign **M** to the **slope** of the regression equation for the well by selecting **M** in the **Select the Variable to Assign** window, toggling on **Constant**, entering the **slope** value in the window, and clicking **Assign**.

9. Click **Next >** to proceed to the **Depth Range Selection** dialog box.

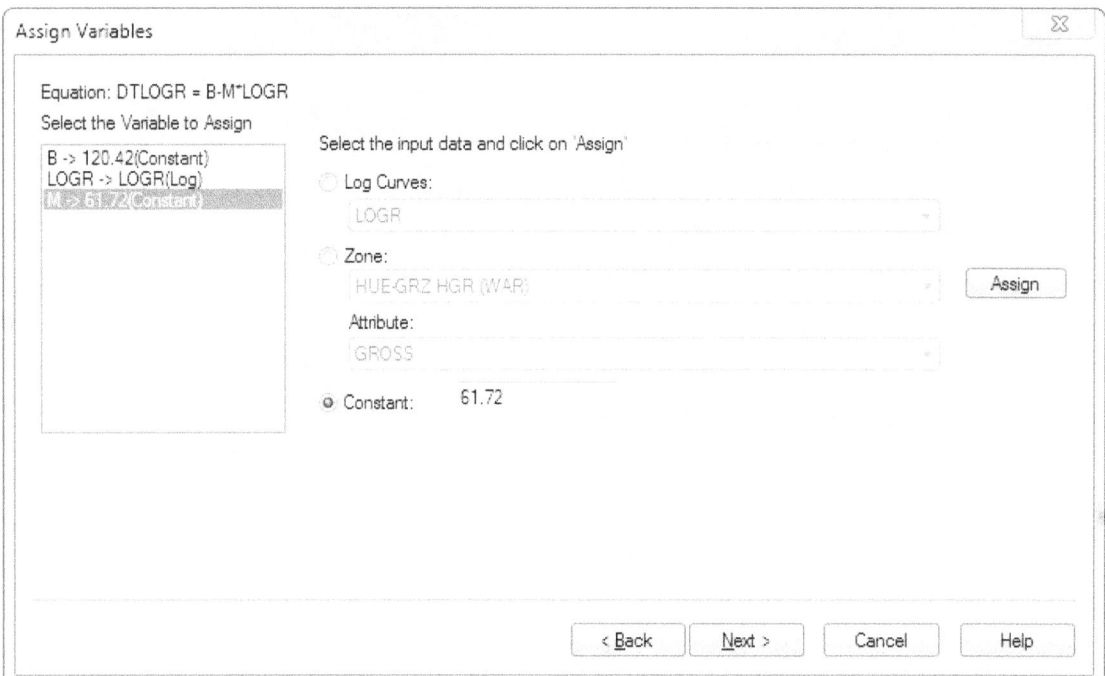

10. Toggle on **by Zone Intervals** and select the **zone to be assessed for probable source rock potential** from the drop-down menu.

11. Click **Next >** to proceed to the **Output** dialog box.

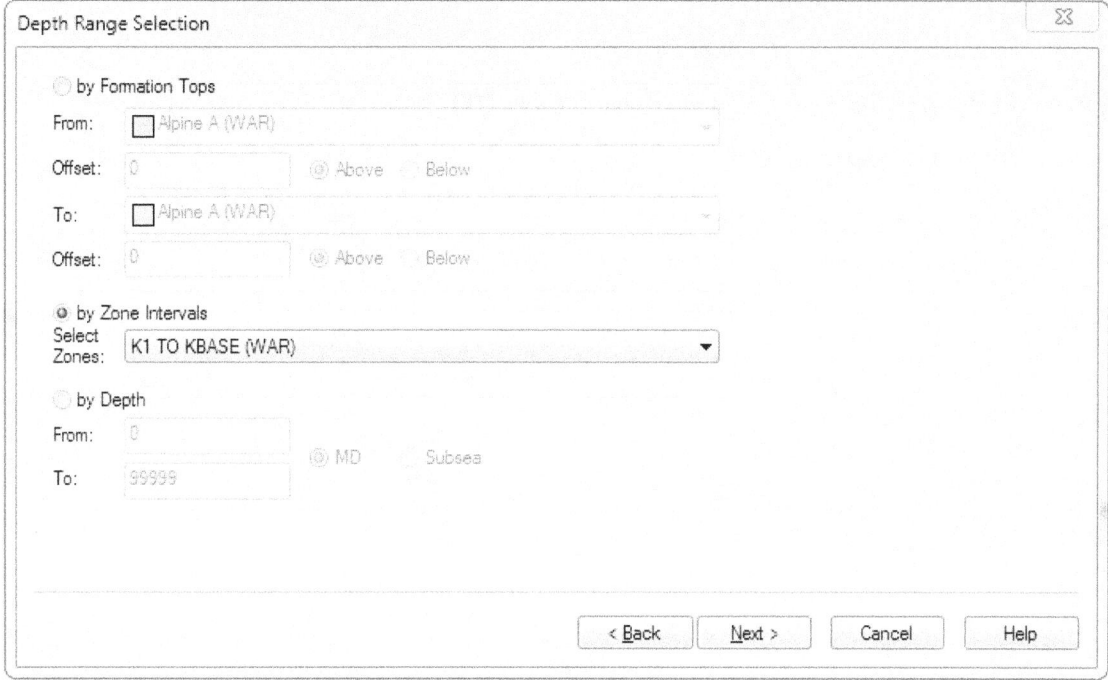

12. In the **Output Log Curve Name** window, type **DTLOGR**.

13. In the **Output Log Curve Type** window, select **Other** from the drop-down menu.

14. Toggle on **Create new log curve and replace existing curve**.

15. Click **Finish**.

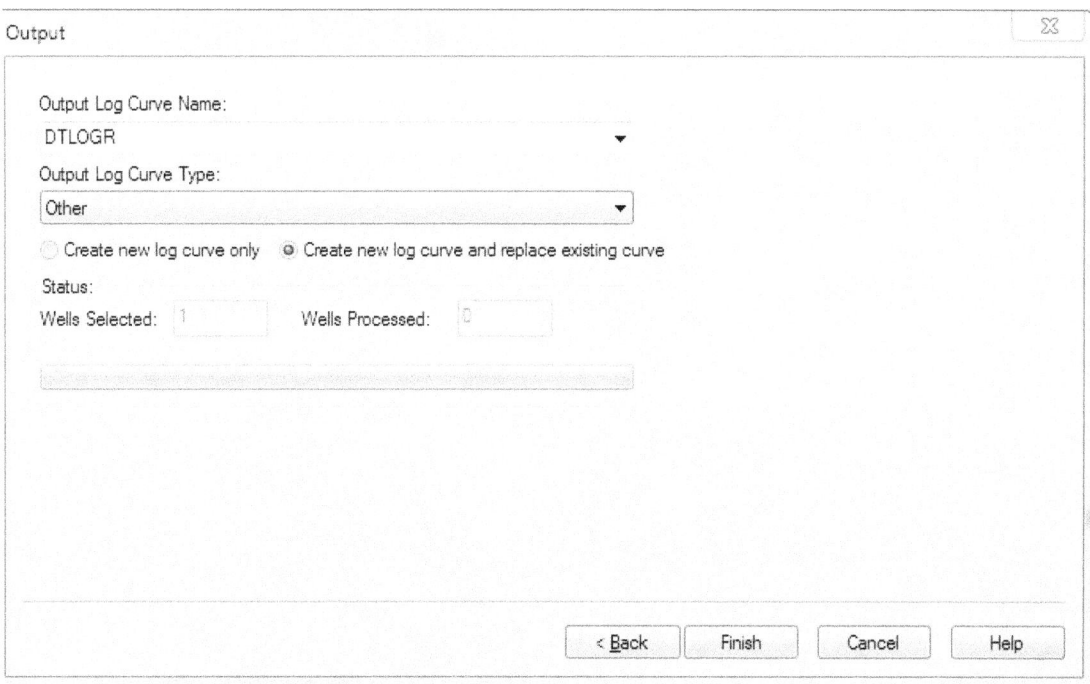

Calculating Δ*DT* Curve Separation

1. From the **Logs** menu, select **Calculations** → **Equation** to proceed to the **Select or Enter an Equation** dialog box.

2. In the **Equation Category** window, type **Other**.

3. In the **Equation** window, type **DELTADT=DT-DTLOGR**.

4. In the **Description** window, type **Curve separation between DT and DTLOGR**.

5. Click **Next** > to proceed to the **Assign Variables** dialog box.

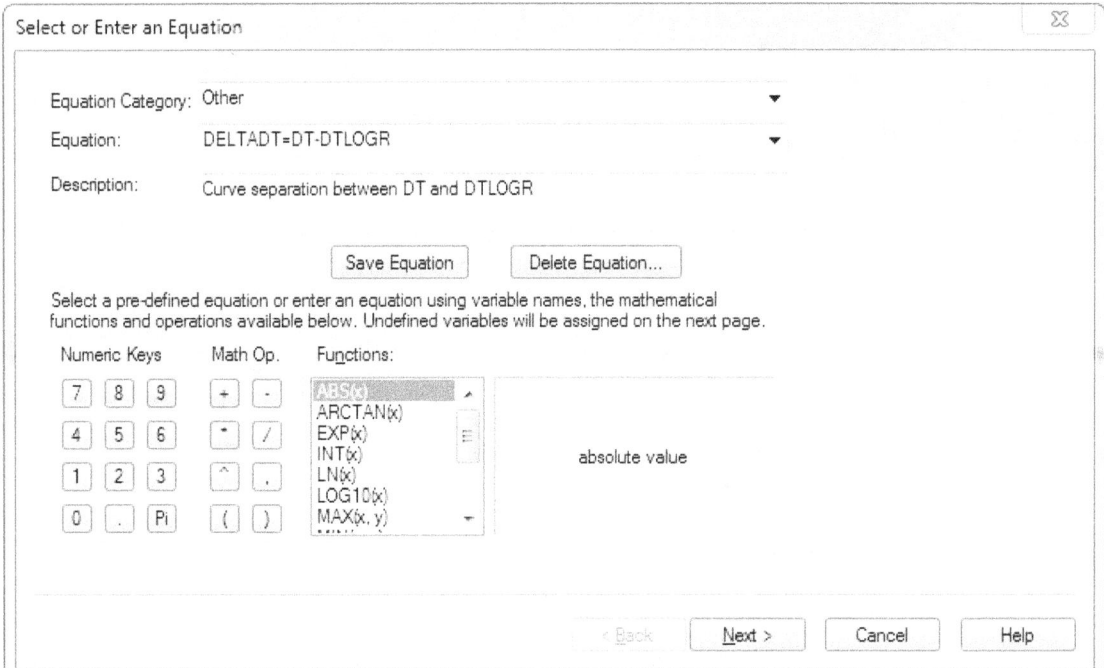

6. Assign **DT** to the **DT** log curve by selecting **DT** in the **Select the Variable to Assign** window, toggling on **Log Curves**, selecting **DT** from the drop-down menu, and clicking **Assign**.

7. Assign **DTLOGR** to the **DTLOGR** log curve by selecting **DTLOGR** in the **Select the Variable to Assign** window, toggling on **Log Curves**, selecting **DTLOGR** from the drop-down menu, and clicking **Assign**.

8. Click **Next >** to proceed to the **Depth Range Selection** dialog box.

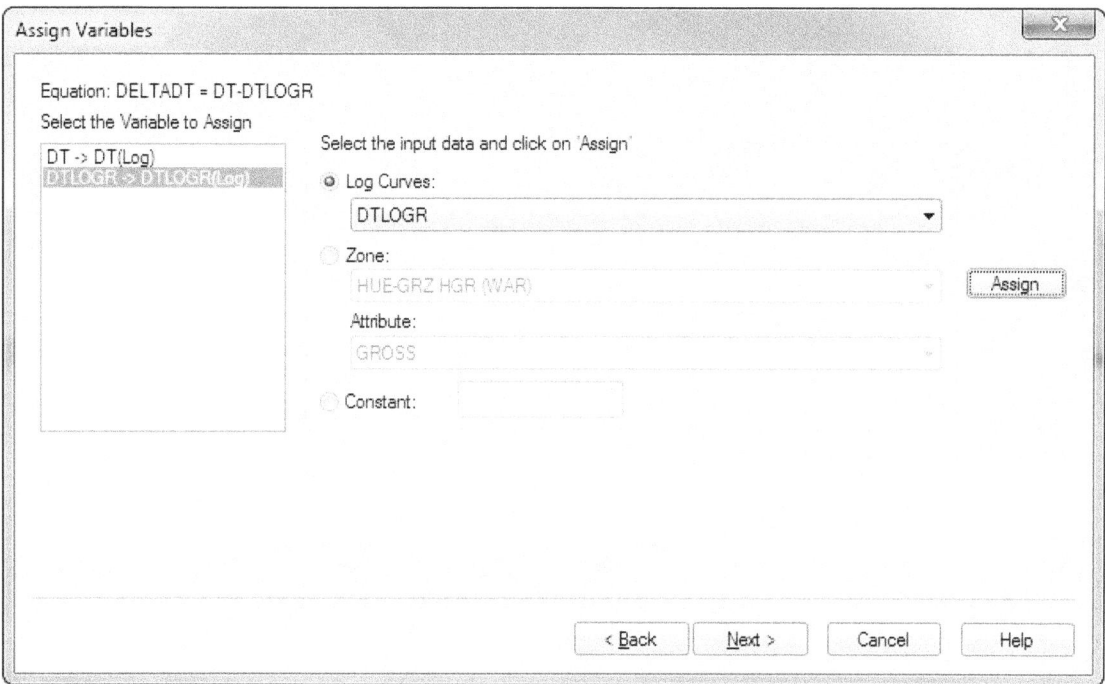

9.   Toggle on **by Zone Intervals** and select the zone to be assessed for probable source rock potential from the drop-down menu.

10.   Click **Next >** to proceed to the **Output** dialog box.

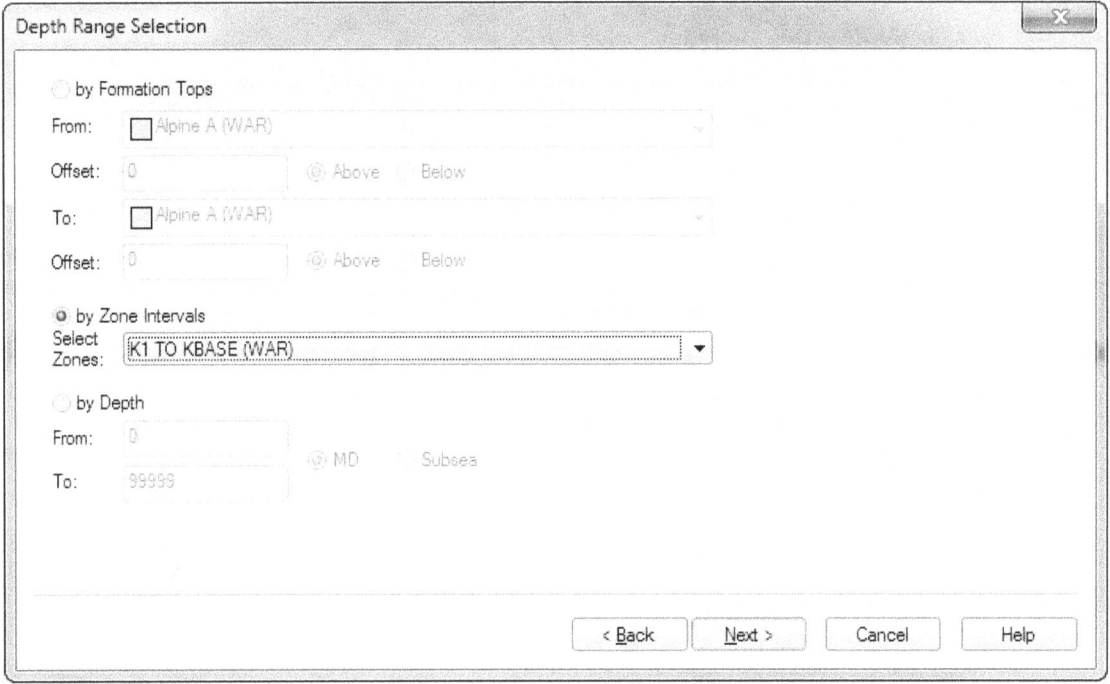

11.   In the **Output Log Curve Name** window, type **DELTADT**.

12.   In the **Output Log Curve Type** window, select **Other** from the drop-down menu.

13.   Toggle on **Create new log curve and replace existing curve**.

14.   Click **Finish**.

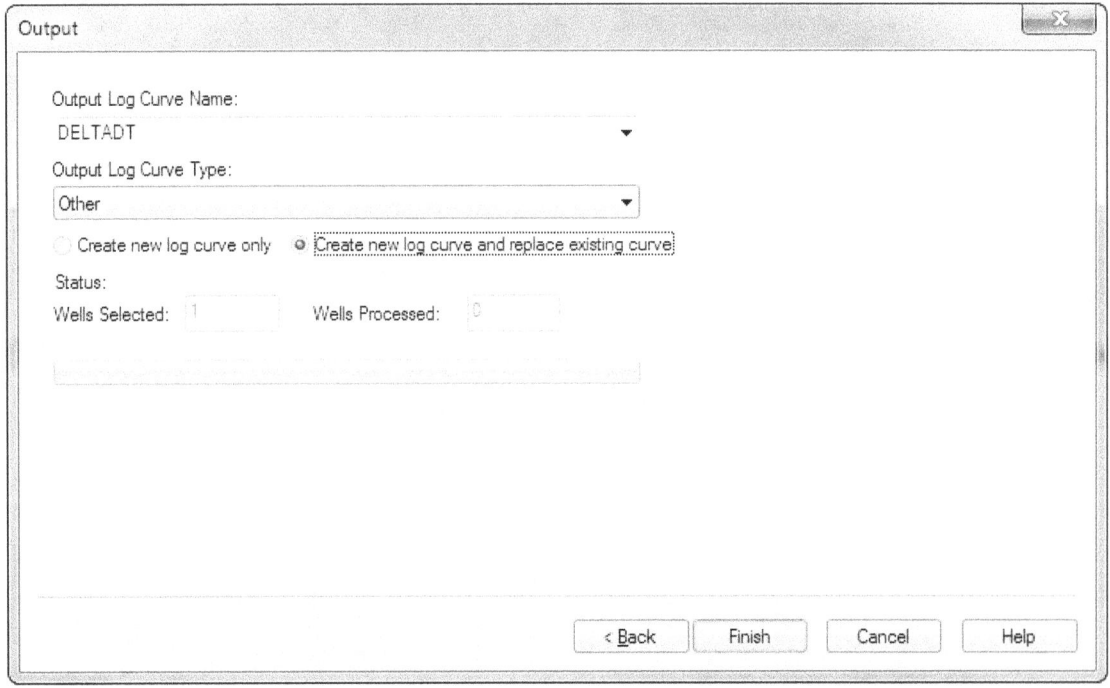

Calculating ΔDT_z

Note: As of this publication, calculation of ΔDT_z entirely in IHS Kingdom® version 8.8 software is not possible. However, the key variables in the equation, and h_{net}, can be obtained using the Kingdom software.

Finding $\Delta DT_{\bar{x}}$

In order to restrict the calculation of ΔDT_z to positive ΔDT values:

1. From the **Logs** menu, select **Calculations** → **If-Then-Else** to proceed to the **If-Then-Else for Log Curves** dialog box.

2. Under **IF**, select **DELTADT** from the **Log Curve Name** drop-down menu, select **GE** (greater than or equal to) from the **is** drop-down menu, toggle on **Constant** and type **0**.

3. Under **THEN**, toggle on **Log Curve Name** and select **DELTADT** from the drop-down menu.

4. Under **ELSE**, toggle on **Missing (NULL)**.

5. Under **Save As**, type **+DELTADT** in the **Log Curve Name** window.

6. Click **OK**.

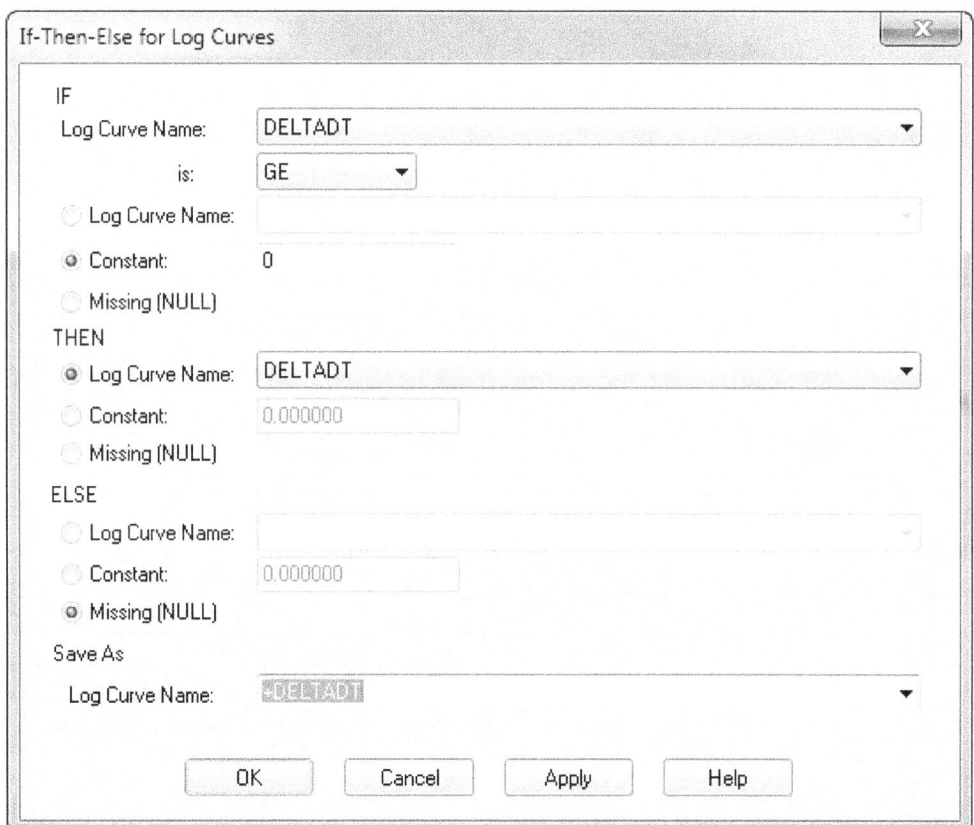

In order to further restrict the calculation of ΔDT_z to intervals containing at least 60 percent shale ($V_{sh} \geq 0.60$):

7. From the **Logs** menu, select **Calculations → If-Then-Else** to proceed to the **If-Then-Else for Log Curves** dialog box.

8. Under **IF**, select **VSH** from the **Log Curve Name** drop-down menu, select **GE** (greater than or equal to) from the **is** drop-down menu, toggle on **Constant** and type **0.6**.

9. Under **THEN**, toggle on **Log Curve Name** and select **+DELTADT** from the drop-down menu.

10. Under **ELSE**, toggle on **Missing (NULL)**.

11. Under **Save As**, type **+DELTADT VSH>0.6** in the **Log Curve Name** window.

12. Click **OK**.

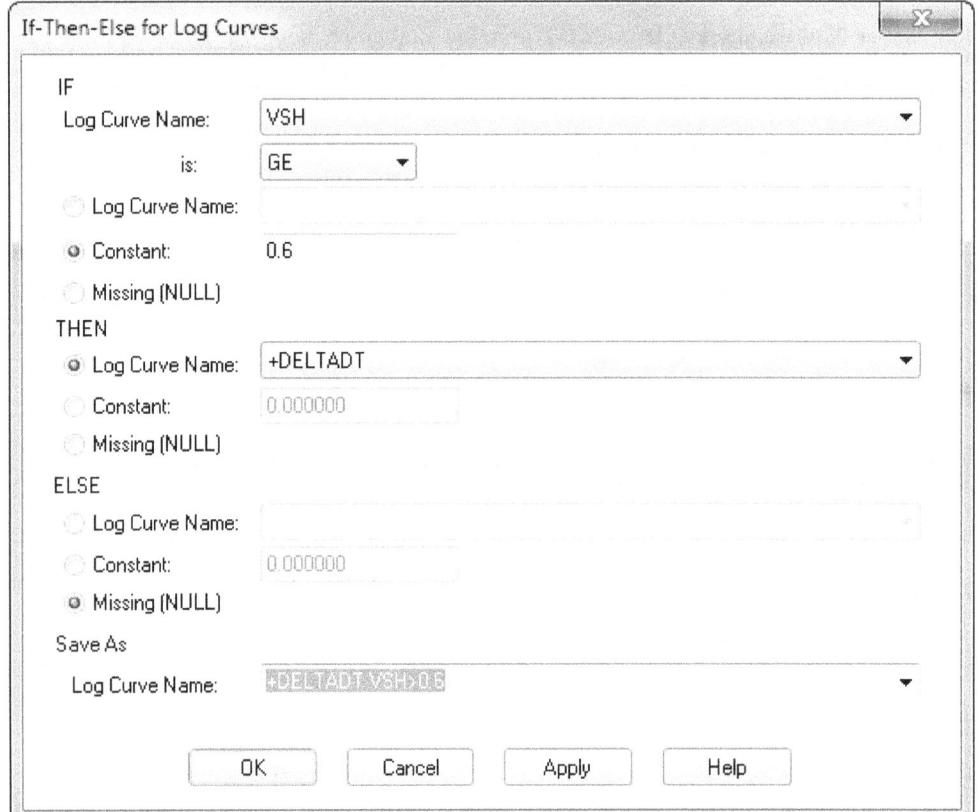

Finally, to find the mean ΔDT value ($\Delta DT_{\bar{x}}$):

13.   From the Logs menu, select **Calculations** → **Simple Statistics** to proceed to the **Log Curve Statistics** dialog box.

14.   Under **Log Curves**, select **+DELTADT VSH>0.6**.

15.   Under **Zones**, select the zone to be assessed for probable source rock potential.

16.   Click the check box next to **Arithmetic Mean (AM)** and type **DDTAM**.

17.   Click **OK**.

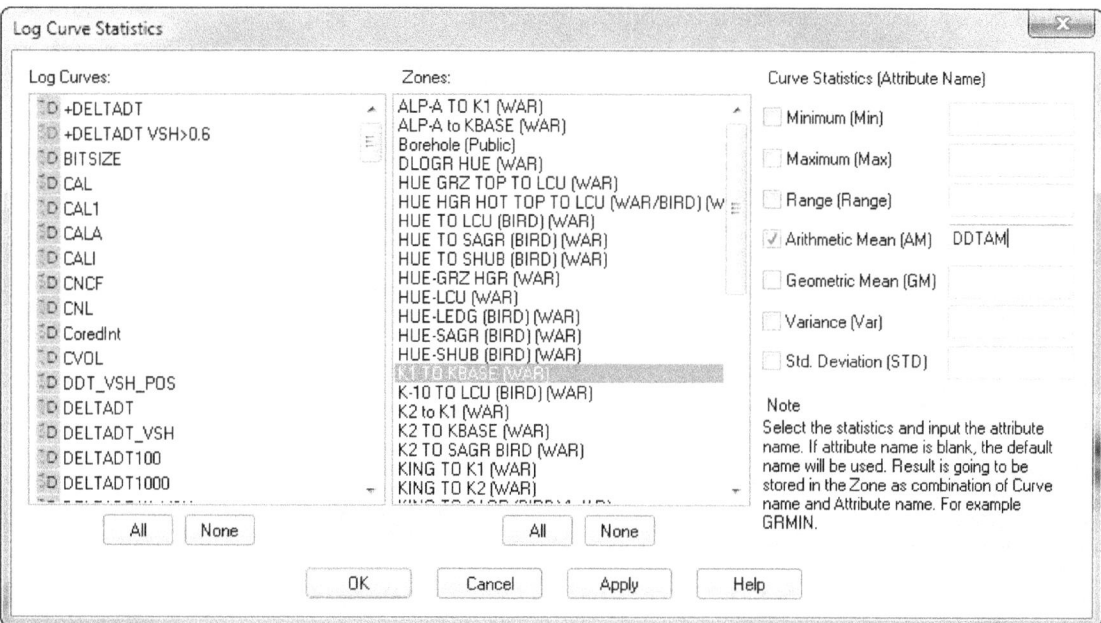

Finding h_{net}

1. From the **Zones** menu, select **Zone Attribute Calculator** to proceed to the **Zone Attribute Calculator** dialog box.

2. In the **Attributes** tab, under **Attributes to Calculate**, check the box next to **Net** and type **HNET** in the window.

3. Check the box next to **Correct for TVD (Elev. Ref.)**.

4. Under **Select One or More Zones**, select the zone to be assessed for probable source rock potential.

5. Under **Conditions**, check the box next to **Other**, select the **+DELTADT VSH>0.6** log curve from the drop-down menu and choose the default values for **Minimum** and **Maximum**.

6. Click **OK**.

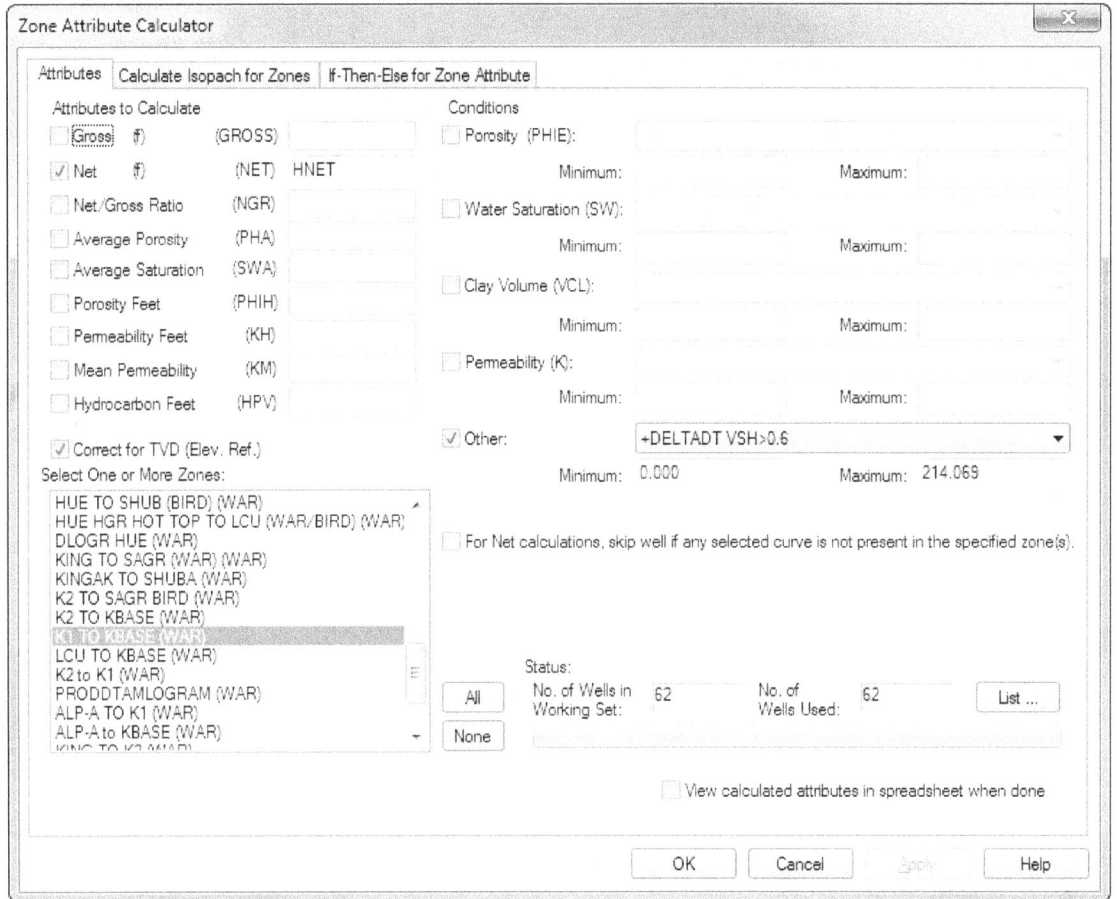

Exporting the Spreadsheet

The user can complete the calculation of ΔDT_z by exporting and h_{net} to an Excel spreadsheet. To accomplish this:

1. From the **Zones** menu, select **View Attribute Values by Zone** to proceed to the **Select Data for Spreadsheet** dialog box.

2. Under **Select Category**, select **Zone Information**.

3. Under **Zones**, select the zone to be assessed for probable source rock potential from the drop-down menu.

4. Under **Attributes**, select **DDTAM** and **HNET** by clicking on each while holding down the 'ctrl' button.

5. Under **Field**, select **Value**.

6. Click > to move your selections to the **Selected Items** window.

7. Click **OK** to proceed to the **Spreadsheet**.

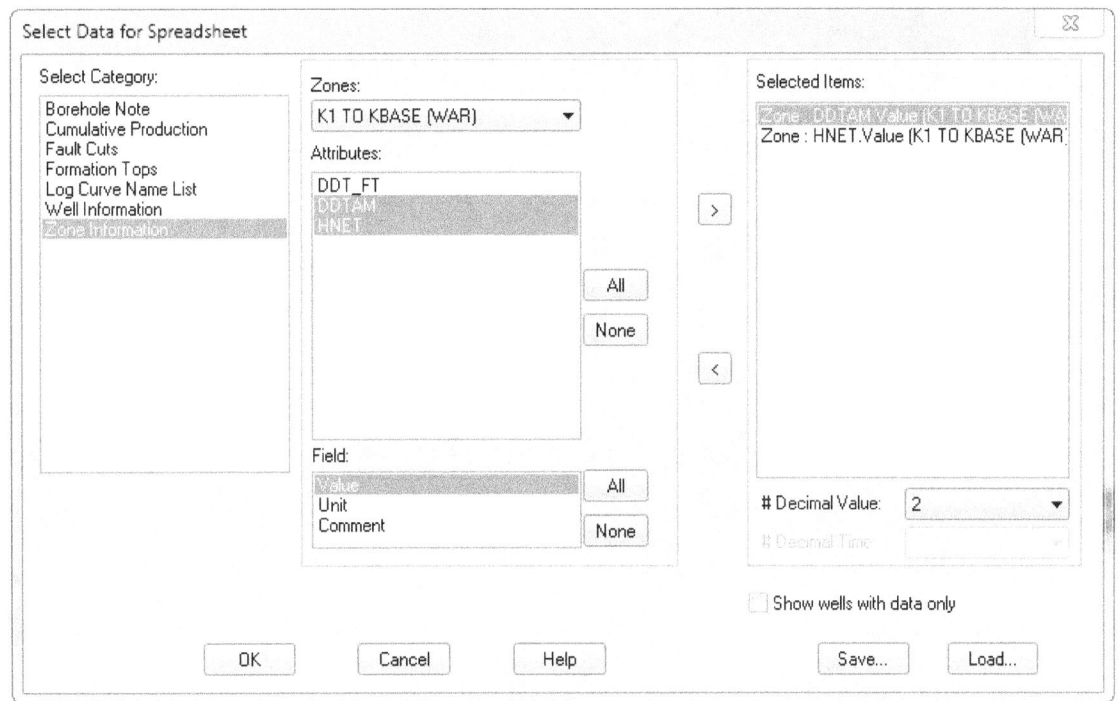

From the spreadsheet, the user can copy and paste the data into an Excel workbook.

	Well Name	Well Numb	Borehole N	UWI	DDTAM.Val	HNET.Value
1	ABEL STATE	1	ABEL STATE	50029202000000	7.60	198.00
2	BEECHY POI	1	BEECHY POI	50029200480000	10.59	181.00
3	BELI UNIT	1	BELI UNIT	50179200020000		
4	BURGLIN	33-1	BURGLIN	50029211060000	17.84	125.00
5	BUSH FEDER	1	BUSH FEDER	50223200040000	19.41	73.00
6	COLVILLE D	1	COLVILLE D	50103200380000	7.93	36.50
7	COLVILLE D	1	COLVILLE D	50103200020000	7.91	15.50
8	COLVILLE S	1	COLVILLE S	50103100020000	7.67	120.00
9	E HARRISON	1	E HARRISON	50703200010000	5.70	30.00
10	EAST UGNU	1	EAST UGNU	50029200520000	7.93	79.00
11	GWYDYR B	1	GWYDYR B	50029203960000	7.94	82.50
12	HEMI SPRING	1	HEMI SPRING	50029210560000	12.89	124.00
13	HEMI SPRING	3	HEMI SPRING	50029212850000	22.98	101.00
14	HIGHLAND S	1	HIGHLAND S	50029201990000	10.80	146.50
15	IKPIKPUK TE	1	IKPIKPUK TE	50279200040000	4.94	157.00
16	INIGOK TEST	1	INIGOK TEST	50279200030000	25.34	235.00
17	JONES ISLA	1	JONES ISLA	50029223190000	5.40	272.50
18	KALUBIK CR	1	KALUBIK CR	50103200010000	9.09	133.00
19	KEMIK	1	KEMIK	50223200060000	15.94	120.00
20	KEMIK UNIT	2	KEMIK UNIT	50223200130000	24.68	23.00
21	KOKODA	5	KOKODA	50279200120000	8.51	170.50
22	KUGRUA TE	1	KUGRUA TE	50163200020000	8.03	605.50
23	KUPARUK	9-11-12	KUPARUK	50029201580000	13.92	118.50
24	KUPARUK S	1	KUPARUK S	50029200080000	24.34	87.50
25	KUYANAK	1	KUYANAK	50163200030000	2.91	58.00
26	LONG ISLAN	1	LONG ISLAN	50029210430000	5.54	76.00
27	MILNE POINT	B-01	MILNE POINT	50029204900000	4.24	58.50
28	MILNE POINT	C-1	MILNE POINT	50029206630000	4.91	102.00
29	MILNE PT UNI	1	MILNE PT UNI	50029203760000	4.49	20.50
30	N KUPARUK	26-12-12	N KUPARUK	50029200320000	7.45	130.50
31	NECHELIK	1	NECHELIK	50103200200000	10.36	234.50

Manuscript approved January 6, 2016

Prepared by the USGS Science Publishing Network

    Publishing Service Centers
    Edited by Natalie Juda and David A. Shields, Reston
    Layout and illustrations by Caryl J. Wipperfurth, Raleigh

For more information concerning this report, contact:

Director, Eastern Energy Resources Science Center
U.S. Geological Survey
12201 Sunrise Valley Drive
Mail Stop 956
Reston, VA 20192

http://energy.usgs.gov/

http://energy.usgs.gov/GeneralInfo/ScienceCenters/Eastern.aspx

www.ingramcontent.com/pod-product-compliance
Lightning Source LLC
Chambersburg PA
CBHW081909170526
45167CB00007B/3212